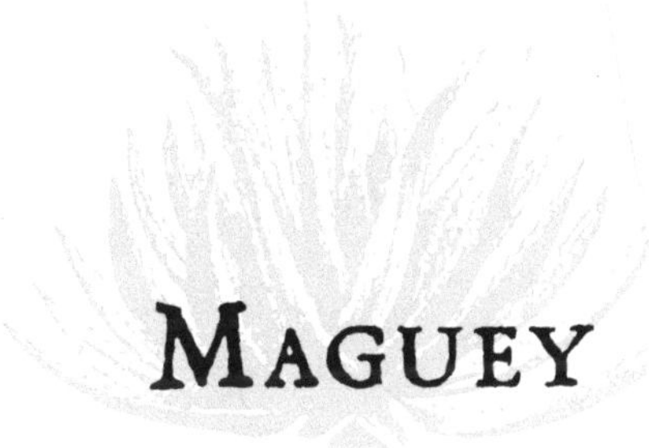

Maguey Journey

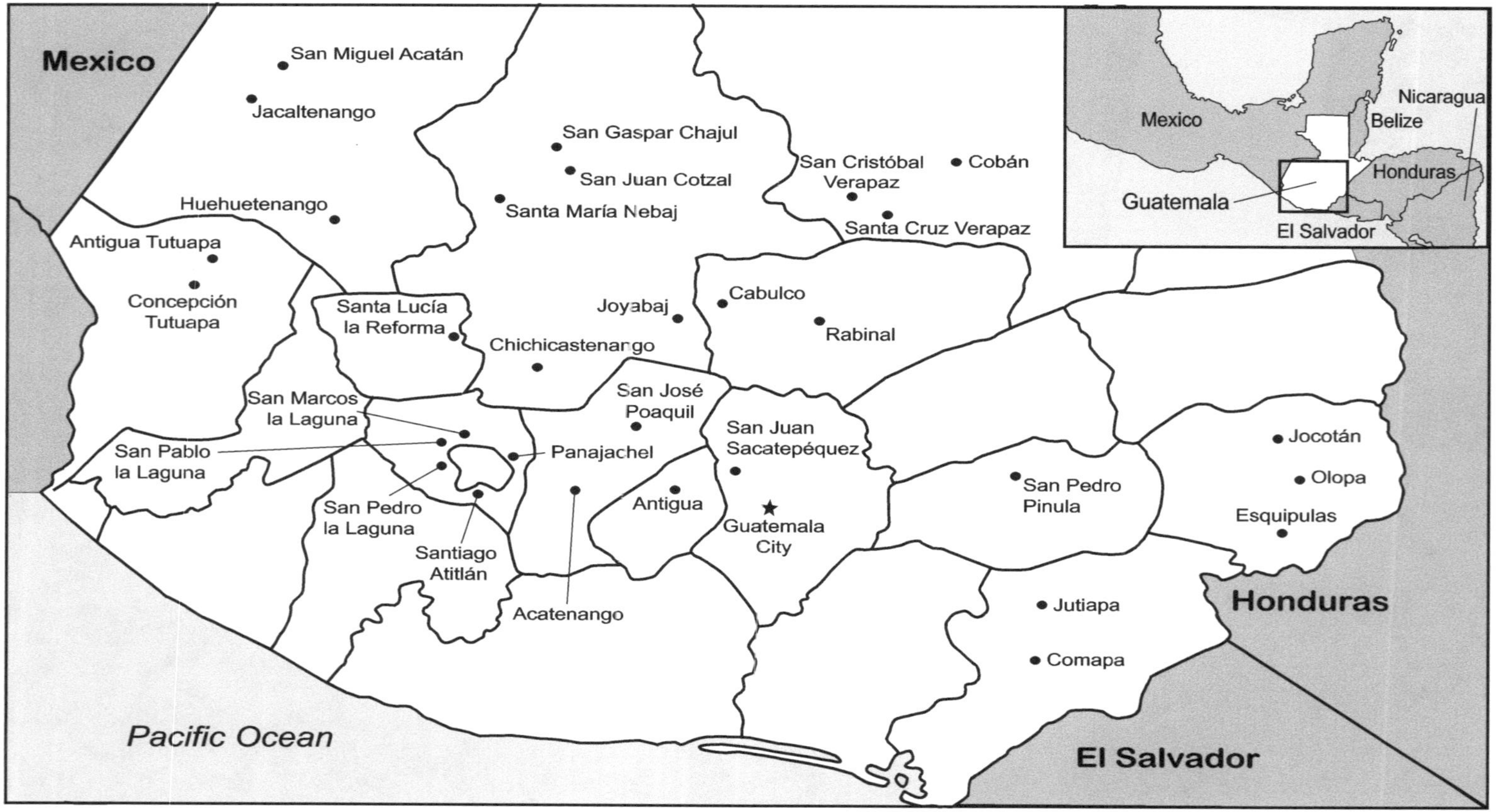

Frontispiece: Primary municipalities in Guatemala visited on the maguey journey. (Drawn by Parisse A. Stewart)

Maguey Journey

Discovering Textiles in Guatemala

Kathryn Rousso

THE UNIVERSITY OF ARIZONA PRESS
Tucson

The University of Arizona Press

www.uapress.arizona.edu

Library of Congress Cataloging-in-Publication Data

Rousso, Kathryn.
Maguey journey : discovering textiles in Guatemala / Kathryn Rousso.
p. cm.
Includes bibliographical references and index.
ISBN 978-0-8165-2698-7 (pbk. : alk. paper)
1. Textile fabrics—Guatemala. 2. Maya textile fabrics—Guatemala. 3. Indian textile fabrics—Guatemala. 4. Hand weaving—Guatemala. 5. Agaves—Guatemala. 6. Agaves—Utilization—Guatemala. I. Title.
F1465.3.T4R675 2010
338.4'76771—dc22
2010011678

Publication of this book was made possible in part by funding from a Fulbright grant.

Manufactured in the United States of America on acid-free, archival-quality paper containing a minimum of 30% post-consumer waste and processed chlorine free.

15 14 13 12 11 10 6 5 4 3 2 1

Dedicated to all of the maguey workers who keep the art alive.

Contents

Illustrations

Figures

Photo Essay following page 126

Preface

Maguey began to play a part in my life when I was a Peace Corps volunteer in Guatemala in the late 1980s. This is because, like everyone else living and working in the *campo* (countryside), I had a *morral* (net bag) in which I carried my daily necessities. At the time I was unaware of how maguey was used to make net bags and related utilitarian objects and, more important, its place as a historical record and visual language. Like most foreigners I focused on the bright, colorful loom-woven textiles, learning backstrap weaving and familiarizing myself with regional clothing styles.

Fifteen years later, after developing my career as a fiber artist, working with natural materials to make hand-constructed textiles, many done in the same technique as net bags, I discovered that little documentation existed on Guatemalan morrales, and I began to conduct fieldwork, to learn what I could firsthand.

Upon my return to Guatemala in 2001, I instantly noticed that many changes had occurred since I had last spent extensive time in the country. One of the most significant events was the 1996 signing of the Peace Accords, which officially ended the thirty-six year civil war. This opened the door to globalization and brought new products, materials, and ideas into even the most remote communities. Also, because of land shortages and low wages, many people have left their homes to earn money elsewhere, which contributes to the decline of many traditions, including maguey. Fortunately, I found enough people still working with maguey that I was able to learn firsthand their techniques and to observe maguey's place in a cultural context.

Maguey textile production, marketing, and transitions are the focus of this book. The first section recounts my journey and experiences, highlighting maguey worker interactions in many locations, and incorporates historical, actual, and non-maguey facts to describe the workers' environments. The plants' journey—the transformation process of turning a raw leaf into beautiful and useful textile products—is next, and, finally, maguey economics as well as many of the transitions taking place. My goal is to illustrate the "state of maguey" at this particular time in history, document the processes, and honor those who keep the art alive.

Acknowledgments

In addition to the maguey workers who welcomed me into their lives, many other people provided guidance and support, and without all of them this book would not exist. Margot Blum Schevill has been a mentor, friend, and inspiration from the beginning. JoAnne Stabb, Gyongy Laky, Barbara Shawcroft, Victoria Rivers, Susan Taber Avilla, Dolph Gotelli, and Emily Dubois expanded my textile education at the University of California, Davis. Librarians Kathleen Wiechelman, at the University of Alaska Southeast, Ketchikan, Alaska, and Michelle Rosensteel, at the College of the Redwoods, Fort Bragg, California, were very helpful. Karen Searle and Wendy C. Hodgson read sections and provided input. Marilyn Anderson, Adele Zang, Mary Jo Turek, Jackie Wollenberg, Janferie Stone, Jill McCaughna, Jim Widess, Anne Wilson, Anne Rock, Ingrid Crickmore, Gary Lybrand, and my parents, Azaria and Saundra Rousso, many of whom read the manuscript during various stages, all gave valuable insight. Parisse A. Stewart drew the map of Guatemala. Thank you also to the University of Arizona Press, especially Allyson Carter.

In Guatemala, the cultural affairs personnel at the United States Embassy provided assistance, as did the curators and staff of the Museo Ixchel del Traje Indígena, in particular Rosario Mirablés de Polanco, Bárbara de Arathoon, Kennis Kim, Anne Girard, and Violeta Gutiérrez Caxaj. Bassilio Estrada Herrera, with Peace Corps Guatemala, contributed information, and Peace Corps volunteers Tahra Taylor, Brett Redwine, Seth Webb, Nicole LaMotte, and Molly Smith helped as well. Those who advised with the scientific names are Mario Veliz, Ana Lucrecia de MacVean, and Abisai García-Mendoza. Susana Heisse is an enthusiastic supporter in San Marcos la Laguna and has written about maguey cultivation and traditions (2000, 2002a–d). Hector René Pena Montes and Celeste Ramírez, of Jocotán, and María Elena Rámoz de Perez, of Olopa, provided assistance. In Antigua, Roberto and Myra Spillari, David and Cynthia Jickling, and Terry Kovick

Biskovich have supported my project for many years. Michael Warren Hopkins assisted with the Spanish translations, and various translators from the K'ulb'il Yol Twitz Paxil, Academia de Lenguas Mayas de Guatemala, provided the Mayan text. In Comapa, the families of Elvia de Jesús Gonzales Tovar, Adela Quiñoles Hernández, and María Elena Quiñoles de Martínez help me out often, as well as Pedro Macario Pérez and Higueros Emilsa Velásquez of Chichicastenango. Debra Rodman and René Benedicto Ruiz Enamorado aided in San Pedro Pinula, and María Fransisca Juárez Morrales in Tutuapa. Thanks also to those in Mexico, El Salvador, Honduras, and Panama who answered my many questions about maguey, bags, baskets, natural dyes, and hats. Finally, the majority of my fieldwork could not have been accomplished without funding from a Fulbright grant.

In addition to my main informants, many others offered advice and information. To give them credit without jeopardizing their privacy I have listed their names, as given to me, in alphabetical order, with no mention of where they live: José Cacho, Rosa Cal, Fransisco Chajil, Margarita Choc Gualim, Fabián Lucas Cor, Domingo (Mingo) Cruz Gómez, Vicente Cuc Caneles, Óscar Fernández, Fidelia Gómez López, Margarita Gómez López, Matilda Gua Lem, Tomás Rámoz Ixtuc, Julia Lassaro, María Margarita Limico Vásquez, Juan López Rodríguez, María Isabelle Macanio Chun, Mariano, Marcos Martiz, Alfonso Morales Girón, Tomasa Morales Raimez, Carmela Morenti, Arminda de Jesús Ohajaca Gonzales, Hipolito Ohajaca Pérez, Gasper Pérez Agular, Alfredo Pérez Galisia, María Pu Castro, Anocleto Ramírez, Santiago Ramírez García, Andrea Rámoz Reynozo, Daniel Rosa, Ángela Sacaria García, Catalina María Sahic, Felicito Sánchez Sánchez, Rejas Sic, Dominga Sicij Sochoj, Diego Us Chic, Juan Us Lux, Gregorio Us Mejía, Eugenio Vacajole, Mariano Vásquez Quisán, Edgar Vásquez Sánsez, and Tino Xinik.

Thank you and gracias to all.

It should be noted that, throughout the text, I adopted the local custom of shortening names, such as Museo Ixchel for Museo Ixchel del Traje Indígena, Chichi for Chichicastenango, and Cotzal for San Juan Cotzal. The spelling of Mayan languages varies and I used those of Nora C. England (2001, 17). The common names for plants, products, and techniques vary widely, and I did my best to use the most common and current ones.

I quoted prices in quetzales (Q), the national currency, and the exchange rate fluctuated between 7.5GTQ (quetzal Guatemalteco) and 7.8GTQ to 1.00USD (U.S. dollar) during the time of my research. Between 2001 and 2007 the daily wage for workers in rural Guatemala was about Q15. Prices quoted can vary, as bargaining is the accepted practice in Guatemala, especially in the markets.

Part I

The Land of Maguey

1

Guatemala

The land of maguey encompasses a region where many similar customs, such as the use of maguey, exist. While the most intense maguey culture has been in central Mexico, Guatemala is another place of importance. Distinguished from neighboring Mexico, Belize, Honduras, and El Salvador by international borders, the republic is made up of many geographical regions, ranging from sea level to over 14,000 feet in elevation. The Altiplano or Occidente (western highlands, between Chimaltenango and the Mexican border), the central highlands (the area surrounding Guatemala City), the Oriente (eastern highlands located between Guatemala City and Honduras and El Salvador), and the Verapazes (a region north of Guatemala City) are where I spent most of my time. I also crossed briefly into Chiapas, Mexico, northwestern El Salvador, and western Honduras to follow cultural, rather than political, divisions.

Guatemala is made up of twenty-two departments, which are separated into municipalities; most are distinct in language, clothing, and customs. These divisions are further broken down into *aldeas* (villages or hamlets) and again into *cantones* or *barrios* (neighborhoods), *caserios* (cluster of homes), and large farms. Spain's colonial influence, beginning in 1523, shaped large towns, which are laid out around a central park or plaza. Bordered by a Catholic church, government offices, and often a health center, market, and school, the plaza is the place where townspeople gather and community activities such as markets and celebrations take place. The biggest fiesta usually occurs on the day of a community's (Catholic) patron saint. Activities always include fireworks, music, dances, processions, elected queens, and markets, and often horse races and rodeos.

In rural areas, small separate sleeping and cooking structures made from the local materials of adobe, wood, or palm leaves are clustered in extended family groupings and surrounded by *milpas* (cornfields). Usually a few chickens, a pig, a couple of dogs, and a cat or two are in the vicinity. Horses and mules are nearby and they are very important as a means of transportation and for carrying products to weekly markets on long, difficult trails through the mountains, because these compounds are often located far from towns or roads. Here people depend on a diet of corn tortillas and black beans.

The rhythm of life is tied to the agricultural cycle. The rainy season usually begins in May, marking the growing season. Throughout this time the milpa is cleaned, fertilized, and maintained until the harvest, which takes place around October, when the rains have stopped. From December until April the fields are cleaned and prepared for the next rainy season. This is also the time when families move to the coast or other regions where landholders of cash crops such as coffee and sugarcane are in need of seasonal workers.

In contrast to the countryside, Guatemala City is a busy, noisy metropolis sitting on a high central plateau. Separated into twenty-one zones, this capital city boasts modern buildings, shopping malls, markets, museums, large parks, an international airport, impressive churches, government buildings, and substantial pollution. Urban and rural bus drivers race through the narrow, crowded streets, charging through each intersection in a game of chance.

While most of my time was spent in rural Guatemala, the journey began in Guatemala City. Commuting on diesel-belching urban buses, I made contact with the appropriate personnel, collected information, and began to plot my movement around the country. When I finally headed away from the capital, on one of the major highways that radiate out from the city, I took a rural, or long-distance, bus.

First-class transports or Pullmans go to large towns, but the majority of the communities that I visited are serviced by second-class transportation, better known as "chicken" buses. These are recycled United States school buses, often with the name of the school district still intact. Most are painted bright colors and affectionately named, such as La Perla del Oriente (Pearl of the East), La Lila de Oxidente (Lilac of the West), Monja Blanca (White Nun, the name of the national flower, one of the orchid family), and Estrella Blanca (White Star). Loud music often blasts out through the windows and the driver will stop anywhere, for anyone, even if the bus is full. Current law states that everyone must be seated and, when police are approached, the driver will motion those standing in the aisle to kneel down, which gives the appearance that all passengers have a spot. Still, there is always room for one more, and on the bus top, for whatever cargo needs to be transported, from sheep to furniture.

An *ayudante* (assistant) yells out destinations, collects fares (somehow remembering who has paid and is owed change), directs passengers, helps in traffic, and loads cargo. Seats built for young students hold entire families of many adults and several children. It is a traveling community, where friends catch up on news, food is sold, and new products are introduced. Salespeople find easy targets from a captive audience and give entertaining lectures on the benefits of herbal medicines, often showing graphic photos of grotesque diseases. English lessons, tools, nutrition, and words of enlightenment are also common sales pitches. Virtual bargains, travel on second-class buses costs only about seven quetzales an hour.

Bus travel is usually slow due to curvy mountain roads, constant stops to pick up or drop off passengers, mechanical breakdowns, and unplanned events such as construction, protests, or procession roadblocks. The meandering pace is often offset by white-knuckle, death-defying spurts of high-speed stunts when passing other slow-moving vehicles. Not surprisingly, there are many accidents, which are highlighted daily in the newspaper with descriptive words and gruesome photos.

Where buses do not travel it is common to ride in the back of pickup trucks. Fares are usually the same as buses and, like the bus, space is always found for everyone and anything that needs a ride. Some trucks are so full it is amazing they do not tip over and can move at all. One driver that I know said he regularly has up to thirty passengers, plus cargo, at a time in the back of his truck. Recent additions to the transportation scene are microbuses, which resemble large vans and hold up to twenty passengers, seen on newly paved roads, and tuk-tuks, three-wheeled motorized mini-taxis, which drivers guide in and out of urban streets in search of potential customers.

Other ways to get around are by bicycle, motorcycle, horseback, or on foot. Walking is often the only option to get from one point to another, and the benefits are numerous. Besides exercise and access to remote locations, many people walk on the roads or trails, and conversations with them often lead to new information. Also, the sounds of the community's inhabitants are better heard when walking and it is easy to follow them to their origin. The noise of maguey leaves hitting rocks as they are thrown down over the person's shoulders can lead one to piles of leaves being processed, and the "whirring" of spinning wheels, where long strands of maguey are being spun, is another clue that maguey workers are active. Both of these activities can be difficult to see when the corn stalks are high. Usually, after I have traveled for a few weeks, a relaxed rhythm is established and the crowded buses, trucks, delays, and other inconveniences become a part of life, mostly enjoyable and certainly memorable.

The locations I selected to conduct my research were determined by my connections with the Peace Corps, the Museo Ixchel, and references to a few publications. I relied on José Balvino Camposeco Mateo's 1994 book for Guatemalan maguey history, some scientific names, and regional usage, although he doesn't cover textile techniques. Marilyn Anderson, Lila O'Neale, and Lilly de Jongh Osborne write about the textiles and crafts, including some that use maguey. Claudia Dary and Aracely Esquivel focus on maguey use in Comapa, Carol Ann Ventura in Jacaltenango, and Samuel K. Lothrop in San Pablo la Laguna. Maguey expert Abisai García-Mendoza identified most of the species listed in this book, and he documents Guatemalan maguey uses in his chapter in the 2006 book *Biodiversidad de Guatemala*. Other botanical information came from the publications of Howard Scott Gentry, Robert W. Schery, and E. G. Squier. Felix Webster McBryde included many maguey facts in his book about southwest Guatemala, and Charles Wisdom wrote about the Ch'orti'. Patricia Rieff Anawalt documented pre-Hispanic clothing, including that made of maguey, and Jeffrey R. and Mary H. Parsons researched maguey in central Mexico and provided "a study of the maguey plant and its role in cultural adaption and cultural change in highland Mesoamerica" (1990, 1). In 2009 I was lucky to obtain a copy of *Ciencia y Técnica Maya*, written by Mónica Berger de White and Matilde Ivic de Monterroso, which contains a very informative chapter on maguey use in San Pablo and San Marcos la Laguna, Sololá.

In addition, I conducted extensive hands-on research, during which I chased rumors, accepted invitations, and followed every lead that came along. In doing so, I filled in gaps and connected many pieces of the maguey puzzle during a pivotal time in Guatemala's history.

My first year in the field I soaked up information like a sponge, and in subsequent years I asked specific questions. Once I was in a maguey area, I found craftspeople as they were selling products in markets, on buses, and via craft organizations, by hotel owner recommendation, from information given by local teachers, and often just by pure luck. Sometimes invitations for observing processes came while walking around the countryside looking for recently harvested plants.

A few times I came across maguey workers by accident in unexpected places, such as the day I took a break from my research and went to Grutas Rey Macros (King Macros Caves) in Alta Verapaz. Upon entering the grounds, I couldn't believe my eyes as a group of six women, all dressed in beautiful white gauze-like *huipiles* (traditional Maya blouses) and blue skirts, were demonstrating maguey spinning and morral making. I spent the entire afternoon with them and, for a day meant to be a rest day, it turned into one of my most productive days that week. Other times, I looked for maguey workers and could not find them, usually when more valuable crops were being harvested. Overall, I saw a broad representation of maguey work around the country and enough repetition in site-specific areas to classify techniques and products.

I should mention that I actively sought out all of the uses for maguey in Guatemala, and I learned that the plant is primarily cultivated for fiber harvesting, as an ornamental, and for live fences. I saw very little evidence of maguey eaten or drunk, with the exception of roasted maguey leaves (imported from Mexico) at the yearly fiesta in Tutuapa. A few people also told me of eating roasted maguey during Semana Santa (Holy Week) in the department of Sololá. García-Mendoza (2006, 182) confirms my findings: Of twenty-two species, eleven are used for fiber; nine for fences; eight for ornamentals; two for food; one for medicine, soap, and miscellaneous; and three for ritual purposes.

The amount of time I spent in each location was based on my own learning curve. Hence, I stayed longer in the first sites than in the latter ones. The majority of my field research was conducted between October 2001 and August 2002 and from January to March 2003. In 2003 I also went to Chiapas, Mexico, to compare maguey traditions there with those of San Andrés Huista, Huehuetenango. In 2009 I traveled to Casa Blanca (an archaeological site and natural dye workshop)in El Salvador, to supplement my knowledge of natural dyes and to learn more about the Pipil people. In addition, I went overland from Guatemala to Panama in 2004, and back to Panama in 2008. These trips gave me a regional perspective and enhanced my work in Guatemala.

In 2007, I traveled for six fast-paced weeks tracking down my primary informants to confirm permission for use of their photos and words in this book. It was an exhausting trip but also valuable, for not only did it give me a chance to ask puzzling new questions in a continuous flow, but I also discovered that, in six years, many things had changed.

With twenty-one Mayan languages spoken in Guatemala, I often hire translators and guides. In addition to being able to communicate with those who do not speak Spanish, this makes it safer when walking around the countryside, as many people believe that foreigners steal children. Once I spend time in an area and the locals get to know my objective, this is less of a problem. In one instance, I was standing at the side of the road waiting for a ride and a woman approached me and said that normally she would fear me but, because of previous "sightings," my purpose was known, and the word on the street was that "the maguey lady" was in town. On the other side of the coin, a few women actually tried to give me their children, possibly because they thought I would take them back to the States and give them a "better" life.

In addition to child-stealing rumors, I am constantly aware of many other issues affecting the country when I travel around. As a by-product of the civil war, there is an abundance of weapons in the countryside, plus a legacy of social violence and a dysfunctional judicial system. Crime is common, and villagers frequently "take the law into their own hands." There are often reports of lynchings and murders in many communities, and two of my sites made the *New York Times* for related events (Gonzáles 2003).

Typically, once I am established in an area, I select two or three people with whom to learn the processes, and because they give me their time, I pay them. I also commission samples of every maguey product found in Guatemala so that I can have examples and see every step of the processes. Beyond observing and absorbing information, I spend hours extracting fibers, spinning them, and making every item to the best of my ability.

I know from experience that items made from natural materials are not beautiful or functional unless the materials are harvested, processed, and woven perfectly each step of the way. Seemingly simple processes and repetitive techniques are actually quite difficult and require years of practice, dedication, and patience. I also know, from basketry studies, the importance of local materials and the connection to ancient lands by the people who use them.

In addition, I have seen that when baskets and bags are considered fine art, their worth usually increases. Baskets and bags are equal in the raw material selection, harvesting, preparation, and techniques and in their potential to be seen as fine art, with the difference being the rigidity and form of the final product.

In pre-colonial Mexico, weavers were admired for their expertise and gained social importance in their communities (Johnson 1971, 311). In Japan, basket artists earn respect and the title of "national treasure" when they reach the highest level of their craft. The same is true in the United States, where those who excel in traditional folk art are recognized as "living cultural treasures." In this spirit, I highlight many recognized maguey workers in Guatemala.

2

People of Maguey

Historically, the people working with and using maguey products were of the lower class. This is widely documented in Anawalt's book in which she writes about clothing in pre-Hispanic Mesoamerica. Cotton, a luxury good, was imported from the coastal region to the highlands and made into upper-class attire, while the commoners wore maguey articles (Anawalt 1981, 12). Even after colonization, Spanish historians wrote that between 1690 and 1699, the *plebeyos* (low-caste Indians), living in large populated centers, tended to use henequén (agave or maguey) fiber to weave their clothing (Osborne 1965, 16). While maguey is no longer made into clothing, most people working with maguey are still very poor.

The necessary skills are learned from generations of family members before them, and toddlers' observation of older siblings and parents is their visual introduction. Many weavers said they began to work full time with maguey around the age of twelve. In most locations, men extract the fibers, women and children spin them, and both men and women create items depending on economic need and tradition. Generally, men make the articles that they use, and regional specialization is common.

Because many of the maguey locations where workers live are very remote, tools are often made from recycled and found objects. Spinning wheels of local wood or used bicycle wheels are common. Long needles are crafted from umbrella ribs, and I even saw an old bullet casing with a protruding bent nail incorporated into a spinning apparatus.

Viewing any or all parts of the maguey process usually requires a great deal of luck and patience as it always comes after cooking, collecting wood, caring for children, and working in the fields. Typically, maguey workers live subsistence lifestyles in areas where abundant cash crops do not grow, and they have few options to earn money. Maguey gives them this opportunity, and when food is scarce or there is an upcoming fiesta, this is often when more people work with maguey.

Even if it is the right time of year and there are no other chores, it isn't easy to find anyone harvesting, spinning, or working with maguey fibers. This is because the fiber extraction is always done outdoors, often in the distant campo, because it is messy, smelly work. People spinning with *taravillas* (rope makers) or wheels are easier to spot, as this is performed outside, in empty streets or fields, closer to home. The making of looped, knit, and linked bags and leg spinning can be done inside or outside, so this activity may be hidden. Loom weaving and hammock making are large projects, worked outside, and are easy to recognize. To add to the uncertainty of locating maguey workers, family compounds are often spread far apart, requiring walking many miles to locate them.

With the exception of the mechanized rope spinning, maguey textile production is a cottage industry, all steps done entirely by hand. Where maguey workers persevere, a mix of tools and techniques is used and a variety of products are made. The following information is based on my interactions with the "people of maguey" in the municipalities mentioned. Throughout my research I saw regional differences in all aspects of maguey use and discovered that, by looking beyond Guatemala's bright fabric rainbow, a whole new textile world opened up to me.

3

The Western Highlands

I began my work in the western highlands, a mountainous region inhabited by several Maya groups. Here colorful *traje* (traditional clothing) is woven on backstrap and treadle looms, and a *corte* (wraparound "skirt") and huipil is the standard woman's attire. In a few areas men wear traje as well, including colorful pants, shirts, hats, and morrales. To get to my first site, the municipality of Concepción Tutuapa, San Marcos, which is near the Mexican border, I took several long bus rides from the capital. Recommended by a curator at the Museo Ixchel, I was armed with letters of introduction from the Museum and the United States Embassy when I went to the weekly mayor's meeting to explain my mission. Fifty-four auxiliary mayors sat on long wooden benches in the town's meeting hall and listened to my proposal in Spanish and the local Mam dialect.

Concepción Tutuapa sits at about 10,000 feet in elevation and has been referred to as "the roof of the world" (Villatoro n.d.). The views are spectacular, including the sight of Central America's highest volcano, Tajamulco, and another, Tacana, which straddles the Mexican border. It is not the easiest place to get to, as the steep, windy roads remain unpaved. Concepción Tutuapa replaced Antigua Tutuapa as the government seat in 1870, about the same time that two Germans settled in the area (Villatoro n.d.). The local rumor is that they didn't cause this change of location, but fears of neighboring Tejutla encroaching on Tutuapa's territory did. The name "Tutuapa" means *comida en la morral* (food in the bag), in Mam.

The women of Concepción Tutuapa wear traje, but it is machine-made and of no specific style, imported from factories near Quetzaltenango and Totonicapán. I was told that the traditional wool traje has not been made or worn for many years. While locally loom-woven traje has disappeared, the backstrap loom is still utilized for weaving small maguey products. The women of a neighboring municipality, Colotenango, continue to weave cotton threads into cloth on backstrap looms.

In this region both men and women use morrales, and women are the primary maguey workers. This is in contrast to other municipalities, where mostly men work with maguey. I was curious about this, and when I asked a man why, he replied without hesitation, "Because morrales have no value [probably referring to economic value]!" But at one time they did, and women were required to master the skills of making one before they could marry. While this is no longer true, many elders remember working with maguey while tending the family's sheep in the countryside.

This large municipality hosts a few weekly markets. I went to them often and saw many women, and a few men, selling maguey products. The most common items sold in Concepción Tutuapa are fibers, *redes* (cargo nets), aniline-dyed morrales, and loom-woven *mecapales* (tumplines) and *cintas* (cinches). Naturally dyed morrales and fibers are sold in the Sochel market, and a variety of maguey objects can be found in the smaller Tutuapa market. Part of the reason for the diversity is that the maguey species utilized in the region grows only at high elevations (above 7,000 feet), and loom-woven product specialization is also common to the higher aldeas, with the naturally dyed items made in the lower ones. If fibers or loom-woven products are sold in Sochel, they most likely came from Concepción for resale.

One early morning in the market, I struck up a conversation with María, who was selling maguey cintas. After I briefed her about my project, she invited me to her house to learn how they were made. When María demonstrated the process, her teenage daughter Suli assisted, while her two young sons watched. At the time, her husband was picking coffee in Mexico and she stayed home with their three youngest children.

On another market day I spoke with a woman named Consuela, whom I have since come to know well. Her adobe home is about forty minutes from "town," and on my first visit I was "lost" for half a day wandering around the beautiful countryside while looking for her house. When I finally found it she introduced me to traditional dyes and spinning. Many of Consuela's friends make fine morrales, and she always seems to have a few on hand to sell when I visit.

I don't remember how or where I met Teresa, but she quietly wove her way into my life. Teresa may have sought me out, as she has been involved in a number of development projects over the years and is willing to try anything new. Teresa lives in the opposite direction from Consuela and has twelve children. Besides making morrales, Teresa sells tamales and hot chocolate in

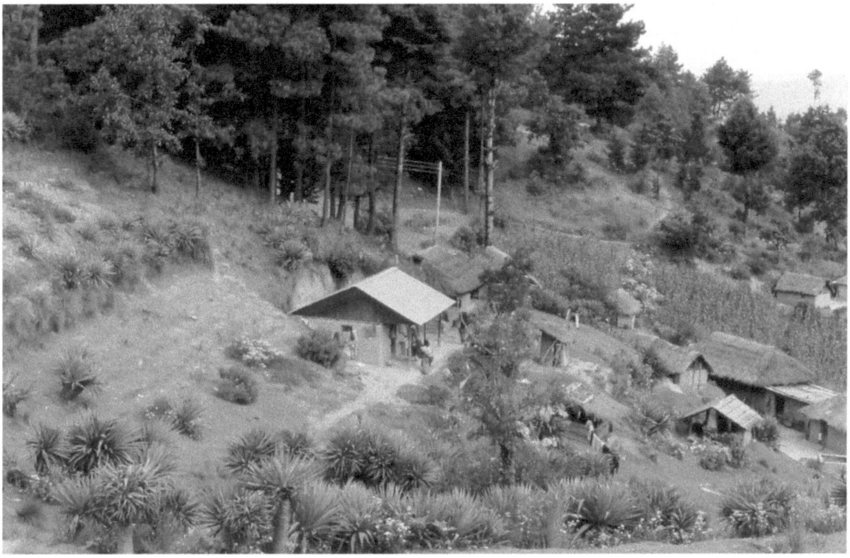

Figure 3.1 Bonifacia's house, where lessons took place, Tutuapa, San Marcos. (Notice the mecate growing nearby.)

the Sunday market. Her daughter-in-law also makes morrales, but is happier crocheting cotton threads than working with maguey strands.

Another woman who recounted her experiences to me was Bonifacia, the sixty-seven-year-old mother of Antigua Tutuapa's then-mayor, Ermiño, whom I met after he and I had drafted a lengthy proposal to attract maguey workers. Bonifacia, her sixteen-year-old granddaughter Elsa, and her eleven-year-old grandson Tránsito were among those who shared their skills with me (fig. 3.1). They introduced me to the fiber-extraction processes and bag-making methods common to this area. Elsa was still learning the finer details and often looked to her grandmother for technical advice, which she did not mind giving. But every time Elsa did not use the tiny bits of maguey left from trimming, Bonifacia scolded her, for when she was a child absolutely nothing was wasted.

During the time I spent in the municipality I was constantly reminded that fewer people worked with maguey compared to *"antes"* (before). Along with wheat and wool, maguey products played a large part in the Concepción Tutuapa economy up until the early 1990s. In the mid-1990s, only about 350 men and 300 women worked full time with maguey (Camposeco Mateo 1994, 50), and by 2001, even fewer. This is in part due to the large number of inhabitants who have left the community to work elsewhere. Plastic fibers have not made inroads as in other regions.

When I left the municipality it was February and, while the days were beginning to warm up, it was nice to leave the cold nights behind. My ride out of town was in the back of a truck to the aldea of Sochel. I went on market day because I knew there would be plenty of transportation to the main highway. Not planning on buying anything, I quickly changed my mind when there were more women than I had ever seen before selling beautiful morrales. This was because the annual fiesta for Santa Candelaria in nearby Chiantla, Jacaltenango, and Huispache was approaching and people needed spending money. After adding a few more morrales to my collection and climbing into the back of another pickup truck, along with my bags, a few other passengers, and their cargo, I headed down the mountain to the warmer climate of the lower valley and a few rest days in Huehuetenango.

From Huehue I went to the municipalities of San Miguel Acatán and Jacaltenango. When I looked at the map it seemed logical that I could combine visits to both in one trip and, after inquiring about travel logistics with the appropriate personnel in the bus terminal, I decided to go to San Miguel first.

The road goes over the Cordillera de las Cuchumatanes, one of the highest mountain ranges in Guatemala. It was paved to the town of Soloma. Upon entering the community the view is dominated by huge new buildings, which, according to my fellow passengers, were built by the local *coyotes* (smugglers who take people to the United States). From Soloma the road became dirt and rapidly deteriorated.

At three-thirty, our estimated arrival time, we were still miles from our destination and the bus abruptly came to a complete stop in front of a disabled truck that was blocking the narrow road. Unable to pass, our driver was soon under the truck and in discussion with the truck driver to determine the problem. With no movement in sight, an hour and a half later a group of us set off by foot. After walking about five miles, we entered the isolated Akateco-speaking mountain town of San Miguel Acatán just before dark.

The next day was market day. After checking it out, I found the maguey co-op mentioned in Camposeco Mateo's book (1994, 49), but it has since become a produce stand. A vendor in the market told me that maguey, once a major industry, is now almost nonexistent. There was one woman buying the few morrales that showed up in the market to resell in Soloma, and colorful, crocheted bags were more common. It seemed that the local maguey economy was just about over, so I went off in search of a ride to Jacaltenango.

I quickly learned that there were no direct roads between the two municipalities and so I had to be come up with a "plan B." I located a pickup truck going to the aldea of Coyo, where I was told there was a foot trail to Vitzupal, one of Jacaltenango's twenty-three Popti'-speaking aldeas. I found, and followed, the route straight down the mountain, and by the time I got to the bottom, my legs were shaking from the steep descent. When I reached the road, a bus departing for Jacaltenango was waiting and I happily jumped aboard.

From Jacaltenango I went to the aldea of San Andrés Huista, which Camposeco Mateo noted as a principal production center (1994, 49). Here a few people still work with maguey, and one of them is Ramón, whom the local teachers taught to make *matates* (loose, openly constructed net bags) twenty years earlier. One of the teachers took me to Ramón's wattle-and-daub (mud-and-stick) house near the school, where he lives with his wife and daughter. I arranged to return the following day to begin documentation, and this led to many additional visits. An observation I realized soon after meeting Ramón was that he is completely blind.

When I returned unannounced in 2007, Ramón was sitting in the exact place I had left him five years before, working on a matate. Three completed bags hung on the wall of his house, one of which held enough prespun fiber to complete the bag he was working on. After catching up on our lives I asked him how matate making was going.

He said it was getting more difficult to obtain fibers for his work because eighty-seven-year-old Andrés, one of the last men to process maguey fibers, was looking to retire. His maguey fields were about a one-hour walk down the mountain, past coffee-tree and sugarcane plantations, but apparently Andrés burned most of his plants in 2006 and has produced very little fiber since. When maguey fibers are no longer available to him, Ramón said that his bag-making days would be over. I mentioned plastic as an alternative and Ramón said he doesn't like the feel of it and therefore won't use it. This is unfortunate as not only does making matates give him something to do, but it also brings in much needed income to his family. When I rode back to Jacaltenango with the teachers, I casually mentioned that a new project, or material, needed to be introduced to Ramón's household, and they agreed. I'm hoping he will still be making matates on my next visit.

The matates that Ramón and a few others make are constructed using the technique of linking, the only place in the entire country where bags are made with this method. This is because their ancestral homeland is a place known as Peb'an (wik'u in Tzohil), located in the state of Chiapas, Mexico, where linked bags are still common and created in the same way.

Legend has it that around a hundred years ago, an unpopular Catholic priest came to visit Peb'an, and the townspeople made him a meal of vultures.

One of the maids felt pity for him and warned the priest about the food, after which he became angry and cursed the town, wishing upon them pests and bats. The townspeople fled in many directions, with some of them going to a place called Corrales Ch'en (Joyas Verdes), but there was insufficient water, and so they eventually moved to their present location (K'ulb'il Yol Twitz Paxil 2001, 52–53).

Once, there was also a healthy maguey industry in Jacaltenango, where the fibers were extracted between two sticks of *caña brava* (bamboo; *Chusquea serrulata*) (Camposeco Mateo 1994, 32, 37; Anderson 1978, 44) and spun on the leg (Ventura 1987, 44; Anderson 1978, 45). Looped and loom-woven bags were common, and I saw a few woven bags and mecapales in the market, but they were all made from synthetic materials.

The general consensus as to why the maguey industry declined in San Andrés Huista and San Miguel Acatán was people's fear of traveling up and down the mountain during the *violencia* (civil war) in the early 1980s. Prior to this, large quantities of maguey fiber were traded by men of Jacaltenango's aldeas to men of San Miguel, who made bags from them (Camposeco Mateo 1994, 49). More recent factors affecting maguey production are the influences of the coffee industry and the large number of men (approximately 65 percent from outlying aldeas) who have migrated to the United States for work. Many inhabitants told me that even the milpas remain unplanted because no one is left to work in them.

From Jacaltenango I took a direct bus back to Huehue, which turned into an exciting ride, for on a steep incline the motor suddenly burst into thick black smoke and flames. The driver managed to stop the bus, get all of the passengers off, and fix the problem all in what seemed like seconds. Afterward, we continued on our way as if nothing had happened and I made it safely to Huehue where I began to plot my next move. This was influenced by prior research in the Phoebe Hearst Museum at the University of California, Berkeley, where I had analyzed maguey bags collected by Gustavus A. Eisen in 1902, some of which were from Santa María Nebaj, El Quiché, where I decided to go next.

In the bus terminal I boarded a bus to Sacapulas, which traveled down through the beautiful Rio Negro valley to the crossroads for Santa Cruz del Quiché, Nebaj, and Cobán. Here, fruit stands line both sides of the pavement, and it can get very hot under the midday sun. After waiting for nearly an hour, I caught a ride in the back of a large truck, with several other people who were heading to Nebaj for the following day's market. They had several large bundles of wares wrapped in colorful cloths, and everyone was in good spirits. We went up and over the mountains and down into the broad valley where Nebaj sits. In 2001, the roads were unpaved and it was a very long, bumpy ride, and rumors of robberies en route made for a tense journey. Since then, the road has been paved, and now microbuses charge up and down the steep grades.

Nebaj, along with the municipalities of San Gaspar Chajul and San Juan Cotzal, is in a region known as the "Ixil Triangle," where similar customs, including the Ixil language, exist. This area was hit very hard during the violencia, when many lives were lost and entire aldeas were destroyed. Located at an elevation of about 6,500 feet, the scenery is again spectacular, with lovely forests, caves, and waterfalls. Coffee, oranges, bananas, peaches, apples, plums, and cherries grow in abundance. Women in all three communities weave beautiful distinct huipiles, *rebozos* (shawls), and *servilletas* (small, general-purpose pieces of fabric used for wrapping items) on backstrap looms. There are also corte factories in Nebaj where yards of red fabric are woven on treadle looms.

The Nebaj traje had shifted slightly in color since my last visit in the 1980s, when it was bright red. The current style is dark purple, and many of the three-piece huipiles now have machine-woven fabric for the outer panels. I found a single maguey morral in the market and the seller, a young girl, told me that it had been her grandfather's. This is very different from what Osborne found in the 1960s, when she wrote: "Cotzal and Nebaj are centers for the bag industry, as well as for string and rope, and every street and every hut has its own workshop. Long fiber twine is twisted from one side of the village street to the other" (1965, 176).

While in Nebaj I learned that palm hats are considered a local handicraft. Baskets are the specialty of Chajul municipality, so important that, at least up until the 1960s, the Dance of the Baskets was performed once a year and the principal dancers wore locally made baskets as ceremonial headgear (Osborne 1965, 197). In addition, I soon discovered that Cotzal is still known for maguey products.

The first time that I traveled from Nebaj to Cotzal it was in one of the most crowded buses I had ever been on. Normally the ayudante collects the bus fare by squeezing through the aisle in the bus, but this time he somehow collected it through the windows from the outside, as there was absolutely no room for him inside. It turned out that coffee harvest was in full swing at nearby Finca San Francisco, and it seemed that the entire population had turned out to pick coffee beans.

While in Cotzal I stayed and ate at the Maguey Hotel and Comedor (restaurant). The owner told me that she named her business in 1998 when the streets were full of women spinning maguey and she saw the importance of this "women's work." Because her kitchen work is also important "women's work," and she didn't want to use her own name, she included "maguey" in the title of her business. A few maguey spinners still work on the streets and I soon befriended them. One, Isabella, buys her fibers from a few men in town and, after spinning, makes *hamacas* (hammocks), which she sells locally and at nearby fiestas.

It just so happened that the yearly Chajul fiesta took place while I was there, and Isabella invited me to go, along with some of her friends. We walked, she barefoot, the two hours to Chajul. Upon arrival, I left the group to check out the activities in town and when I returned a short time later, they were gone, for the hammocks she brought to sell must have sold quickly. I retraced my steps to the church and spent most of the day absorbing the colorful traje from many regions, the beautiful flowers, and the warmth of at least a thousand candles. This is a pilgrimage site for many in the western highlands and, in the words of a gringo I met, "the most traditional town in Guatemala that a road goes to." Scheduled on the second Friday of Lent, Chajul's patron saint is Christ of Golgotha (Kelsey and Osborne 1961, 51).

A man I encountered in the church reminisced about his previous eighteen pilgrimages to Chajul, three of which he walked for three days, each way. Popular souvenirs for many of the visitors are vine baskets of all sizes, made by men especially for the festival. Maguey redes are also sold, and one man in the market had a very nice morral but replied that "it already had an owner" when I tried to buy it.

In Cotzal, maguey is typically grown in rows near fire pits, scattered between the aldeas of Tuxivila, San Felipe Chenla, Tixelap, Santa Averlina, and Chichel. Usually a group of men will go out for a week to process the fibers. This mirrors Parsons and Parsons' findings in central Mexico—they concluded that fiber processing is more efficient when it is a group effort (1990, 299). Sometimes, however, there is no choice, and one day I observed a single man scraping leaves in a remote field. One man said that Cotzal maguey is far superior to that of the Cobán area, although he wasn't sure exactly why. One of the men who demonstrated the fiber-extraction process was Nicolás, whom the president of Asociación de Desarrollo Integral Tetz K'usal recommended. Nicolás has over one hundred maguey plants located about a two-hour walk from town, but as he wanted to work close to home for the demonstration, we harvested a single plant in his backyard, in an aldea about a half-hour's walk from Cotzal. Nicolás and his father-in-law make bags with the fibers, as do many others, whom I observed during my time in the area.

Another bag maker was a seventy-year-old man living near the hotel. He walked with a severe stoop, which neighbors informed me was due to his younger years spent scraping maguey. During the time I was in Cotzal, he worked daily on a figure-eight-style matate and told me the bag was going to have a total of twenty-two rows, each row taking him an hour to complete. The entire bag would be made with fiber from about ten maguey leaves and sell for approximately forty quetzales. When I returned in 2007 he had passed away. That same year, I was pleasantly surprised to see a new sign at the entrance of town that read "Bienvenidos a San Juan Cotzal, Tierra del Maguey, Administración 2004–2008" (Welcome to San Juan Cotzal, Land of Maguey, Administration 2004–2008).

In the early 1990s, the demand for maguey fiber was so high in Cotzal that it was imported from Santa Lucía la Reforma, Totonicapán. However, some in Cotzal felt that the fibers were too soft and so they did not like to use them. Curiosity about this led me to Santa Lucía's maguey aldeas of Ichomchaj and Gualtux, which are located in the high, dry, pine-covered plains west of Santa Cruz del Quiché. This is the center of K'iche' territory, the largest linguistic group in Guatemala and one of the dominant tribes before the Spanish conquest.

Nearby, Santa María Chiquimula, Totonicapán, and San Antonio Ilotenango, El Quiché, are also maguey-producing areas with processes identical to those of Santa Lucía. I visited a rope maker in San Antonio, but most of his work was with plastic, and the small amount of maguey that he used was imported from Costa Rica.

Many Cotzal bags are sold in Chichicastenango, another Ki'che'-speaking municipality that hosts one of the biggest tourist markets in the country. On Thursdays and Sundays vendors come from all over the western highlands to sell their wares. Tourists arrive by the busload to find bargains and take photos of the colorful scene. There are two men who sell morrales in the market, and one of them, Tomas, I got to know well.

He inherited the market-stall location from his father and has been selling bags in the same space since 1984. The bag business is in the inner sanctuary, where mostly locals shop, although an occasional tourist sometimes wanders by. Within his booth, bags from Cotzal, Jocotán, and Concepción Tutuapa, and locally made knit maguey, cotton, and wool bags, strategically hang from bamboo rods and are piled on wooden slabs (fig. 3.2). Tomas also has a small photo album with many photos of maguey fiber extraction and sheep-wool spinning, which he proudly showed me.

From Chichi I traveled by car to San Pablo la Laguna, Sololá, a town described in Samuel K. Lothrop's 1929 maguey article. I stayed in nearby

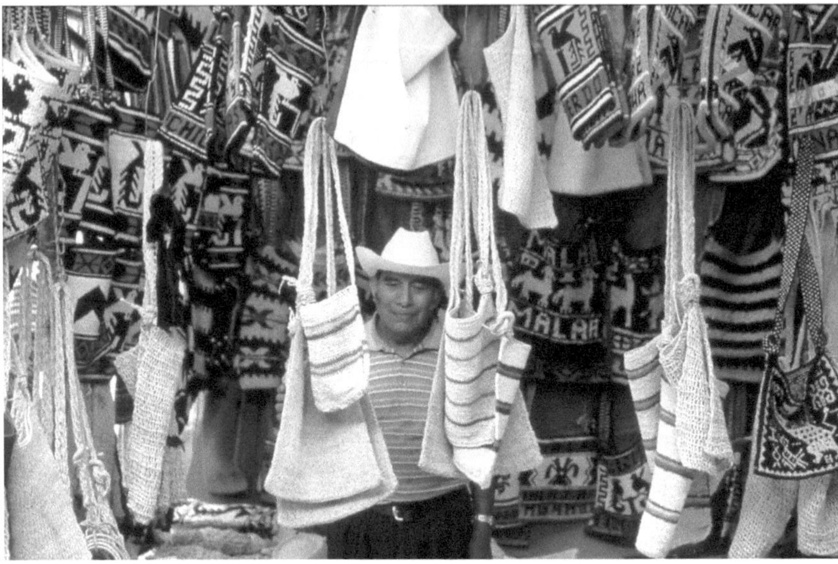

Figure 3.2 Tomas and his market stall where he sells morrales, Chichicastenango, El Quiché. The ones in front are maguey and those behind are wool and cotton.

San Marcos la Laguna because there were more guesthouses and, while I was there, I discovered a small maguey industry as well. While San Pablo and San Marcos are only a couple of miles apart, the people of San Marcos speak Kaqchikel, and those of San Pablo speak Tz'utujil.

The two municipalities sit at about 5,200 feet in elevation on the northwestern shore of Lake Atitlán, where steep mountain ridges rise out of the deep blue waters. Neither has aldeas or weekly markets, with the closest markets in Panajachel, Santiago Atitlán, and San Pedro la Laguna. The entire area is popular with foreigners and, along with the tourism industry, loom-woven textiles, coffee, fish, crabs, and *petates* (woven mats) made from *tul* (lake reeds; *Typha domingensis*)(MacVean 2006, 160) are sold. Along the lakeshore it is common to see bundles of drying tul, and in San Marcos women weave red yardage on backstrap looms, which is then fashioned into huipiles with colorful embroidery around the neck; these blouses are worn with a blue corte and accompanied by a *tzute* (two-paneled square piece of cloth) that is white with colored side stripes. In San Pablo, young girls sit in their doorways and crochet colorful cotton thread into Hacky Sacks (footbags), which are marketed in Panajachel.

During the coffee harvest, buyers with their scales strategically position themselves along the backroads, waiting for incoming *costales* (sacks) of freshly picked coffee beans. The men are from San Pedro la Laguna, where six *beneficios* (coffee associations) are located. Coffee was introduced to the west side of Lake Atitlán in the 1960s, which is also a time when coffee consumption peaked in the United States (Museo de Café 2007). Prior to coffee, maguey was the major crop in the area, and a few men I talked to remembered when they tended maguey plants instead of coffee trees.

Monetary income has always been important in this area because of a lack of abundant flat, fertile land. In the early to middle 1900s, when men found it necessary to supplement their agricultural subsistence, they made maguey rope. Lothrop noticed this and wrote: "For the people of San Pablo the making of rope and twine is practically their only occupation" (1929, 128). O'Neale noted this as well (1945, 20).

McBryde also mentions the rope-making industry of San Pedro, where, in the 1940s, he saw *sogas* (large-sized ropes, about five-eighths of an inch wide and twenty-two feet long) being spun. During that time, he observed that the best maguey grew near San Pedro la Laguna, and he learned that native species were used along with cultivated species for at least fifty years prior to his visit. McBryde also observed the men of San Pablo attempting to obtain superior maguey by transplanting shoots from San Pedro, but they were unsuccessful, perhaps, he notes, due to soil differences. Because the fibers were better from the San Pedro plants, men from San Pablo would often purchase maguey leaves and transport them back to San Pablo in canoes (McBryde 1945, 103). Twenty years later, Nash mentions men of San Pablo importing maguey from San Pedro (1967, 93).

When I visited, beginning in 2002, the small amount of maguey fiber used in San Pedro came from approximately seven men working full time with maguey in San Marcos and San Pablo. One of the men told me that, in the 1990s, many maguey plants were pulled up or burned to create space for planting coffee. But there was still a demand for maguey in the area and so workers in San Pablo rented maguey plants from those in San Marcos for about four quetzales for every hundred leaves cut. In spite of the maguey industry's decline, a woman recently claimed that San Pablo is still known as "the land of maguey" (Berger de White and Ivic de Monterrosa 2008, 136).

In San Pablo, people often spin maguey fibers with taravillas in the streets. This is because a long, straight, unobstructed work area allows for easy spinning. One day while I was observing this, many passersby recommended that I visit Pablo, at his small adobe house located near the top of the community. On my first visit, and every one thereafter, I would find him and his wife, María, spinning maguey fibers bought from several men in town. They make rope, redes, and morrales to sell in Santiago Atitlán.

In both San Pablo and San Marcos, men walk through town loaded down with trimmed leaves on their backs, held in place with a mecapal (fig. 3.3). The trek is necessary because the maguey plants grow high up on the steep hillsides, far from the fiber processing sites, which are located below town, next to the river and lake. This was not always so, as O'Neale wrote of maguey plants growing near the water (1945, 20), but coffee and vegetables grow closest to the lake now.

One man who demonstrated fiber extraction to me in San Marcos was seventy-two-year-old Lucas, whom I had met while standing in the back of a pickup truck en route from San Pedro to San Marcos. We were packed like sardines, along with many other passengers, baskets of chickens, a pig, and a dog, when I explained my project. He mentioned that he just happened to have a few leaves soaking in the lake near the public boat dock, which is one of two traditional processing sites in San Marcos. He invited me to visit him there the next day.

Besides fiber extraction, Lucas taught me how to make a bag and lessons took place at his house, on top of the hill, where I made many trips, usually early or late in the day, when it was cool. It was here that I would find him seated on a wood chair spinning maguey on his leg or making a morral using a method I call "finger-knitting," in which the worker's fingers move the working strand between three tapered sticks, instead of using pointed sticks or needles, such as with traditional knitting. The men of San Marcos make fine bags, and so they need a softer fiber, which is achieved by a slightly different fiber-extraction process than that used in San Pablo.

One day I wandered through town and saw Lucas working while sitting just inside the front doors of the huge new stone Catholic church, as it was his family's day to "guard" the holy building. Lucas was one of the last men to wear the traditional dress of short white pants and a brown wool *rodillera* (blanket kilt) in San Marcos, and it was a great sight to see him seated in the church doorway, working on a maguey morral. Lucas passed away in 2005, a year after his neighbor Francisco, another morral maker, had died. For years, Lucas provided Francisco with maguey fibers, which he knit into the finest morrales I saw in Guatemala (fig. 3.4). So far, their tradition is continuing, with one of Lucas's sons providing fibers to Francisco's son-in-law, who continues to make very fine morrales.

From San Marcos I went by *lancha* (small motorboat) to Panajachel. These passenger boats are often overloaded, have few life jackets, and are very unstable if the lake is rough. Luckily, on the day that I left, the

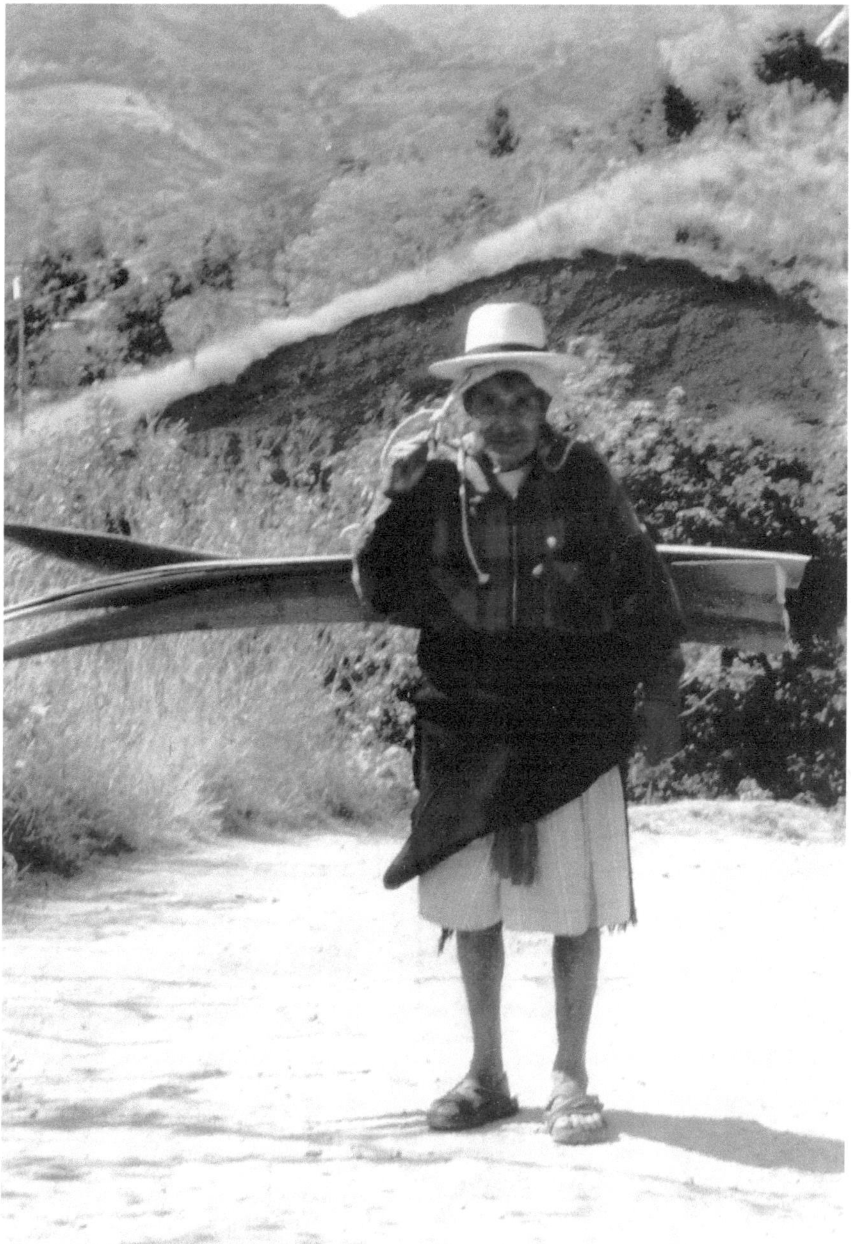

Figure 3.3 Lucas using a mecapal to carry maguey leaves from the field to the processing site, San Marcos la Laguna, Sololá.

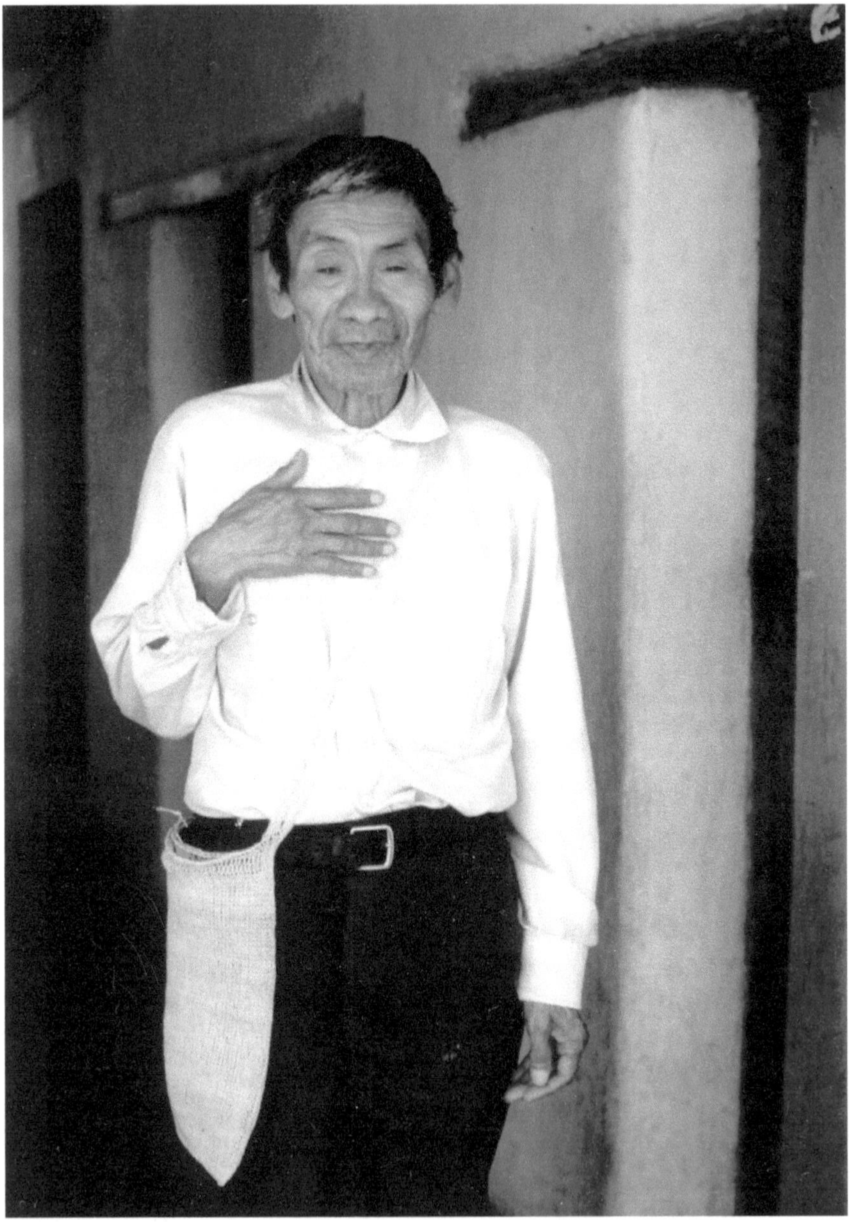

Figure 3.4 Francisco wearing a fine morral that he made, San Marcos la Laguna, Sololá.

water was flat calm and few passengers were aboard. While in Panajachel, I combed the market stalls looking for maguey products until I found a bag that was very overpriced. I asked why and the vendor told me that the person made it with his feet, which I thought highly unlikely, and so I did not bargain any further and moved on.

4

The Central Highlands

The central highlands surround Guatemala City, and I am most familiar with this area because it is where I spent two years as a Peace Corps volunteer. Slightly lower than the western region, the term "eternal spring" is often used in reference to this zone as most of the year the climate is considered perfect. Several volcanoes loom over large fertile plains, and much of the land is cultivated with export-quality produce and coffee. Kaqchikel, the second-largest Maya language group in the country, is spoken in much of this region, and weavers create many beautiful textile items. During the time that I explored the central area, I rented an apartment in Antigua and made short trips to the surrounding maguey communities.

The first municipality I visited was Acatenango, Chimaltenango, a few bus rides away. I went there because Osborne wrote that "the best quality of *pita floja* [untwisted agave fiber] grows in Acatenango and the people of this municipality claim that their product is better than that of San Pablo la Laguna because of the way in which the fiber is separated from the pulp" (1965, 175). Her visit occurred forty years before mine and, when I got off the bus in 2002, I did not see any maguey plants, only the shiny green leaves of coffee. Knowing of maguey's history, I went door to door until I found sixty-year-old Eugenio, in a nearby aldea, where he lives in a small adobe house with his wife, Christina, their three children, and eight grandchildren.

Orphaned as a child, Eugenio's uncle took him under his wing and taught him the details of the maguey industry when he was young. He spent most of his life working with maguey and told me that, during the 1970s and 1980s, everyone in his community was involved with the maguey industry in one way or another. By the time I paid a visit, he had one maguey plant left and, with that, he showed me the extraction process.

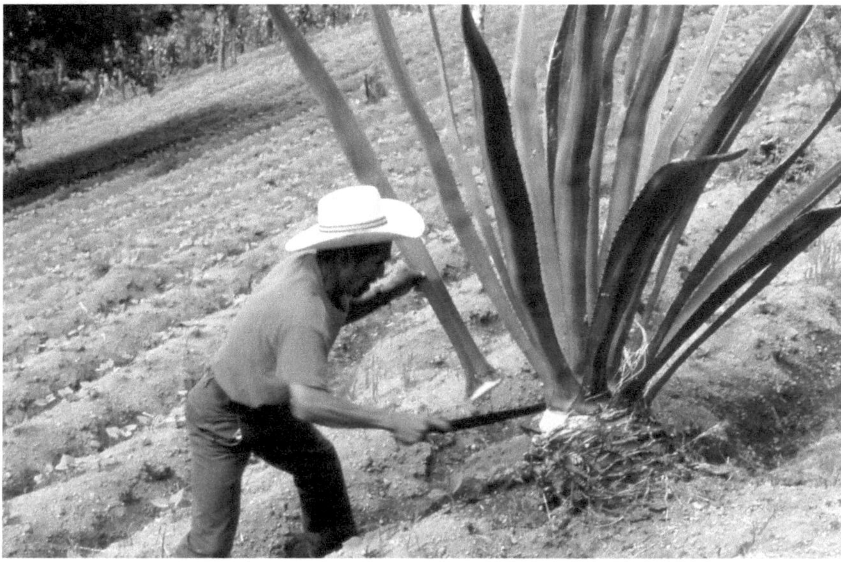

Figure 4.1 Eugenio and his single maguey plant, Acatenango, Chimaltenango.

The plant sat in the middle of a newly planted cornfield on a steep hillside above town and was used by several family members (fig. 4.1). When we arrived, there was a pile of trimmed maguey leaves near the plant that someone else had recently cut. Once the demonstration was complete, Eugenio showed me his pride and joy, the family's coffee plantation. After wandering through fields of coffee, tomatoes, peppers, and corn, we returned to his home, where he continued to tell me about his maguey work.

When I arrived at Eugenio's house in 2007, it was pure luck that he was there, as his entire family had departed at two in the morning on a chartered bus to Esquipulas (located near the border of El Salvador, this site is considered by some to be the "center of the world," and thousands from all over the region make pilgrimages to see the black Christ housed in the massive basilica). He stayed behind to "guard" the house and, to keep busy, he worked on a maguey matate.

There was a pile of maguey fibers nearby. It happened that the single plant we had visited in 2002 had died, and three *hijos* (sons, or new shoots) were growing in its place. Before the "mother" plant became unusable, he had harvested all of the leaves, because he knew that the hijos wouldn't mature for a couple of years. He estimated that the fiber extracted from approximately 150 leaves was enough for about ten matates. I inquired as to the influx of plastic fibers and if he would consider using them in place of maguey, but he said that "they don't spin as tight as maguey fibers," and so he doesn't like them and therefore won't use them.

During my last visit he took me to see a woman who makes petates, a mat commonly woven with rush (*Cyperus canus* Presl) purchased from nearby Yecapa. Leaves are softened in water and split into three parts. The pithy middle is used to make *sopladores* (fire fans), while the other two sections are woven into petates. After the petate demonstration I walked to the bus stop. While waiting for the bus, I couldn't help looking toward the flanks of Volcán Acatenango and wondering what the landscape must have looked like when maguey, instead of coffee, was king.

A stunning maguey *alforja* (saddlebag), which resembles one that O'Neale documents (1945, 182 and fig. 60) in the Museo Ixchel's collection, led me to San Juan Sacatepéquez, known for flower production and the inhabitants' distinct traje. When the bus dropped me at the plaza, I was suddenly surrounded by women wearing purple and yellow striped huipiles and selling colorful flowers. At an elevation of about 5,000 feet, San Juan has twenty-seven aldeas, which are strung out along a paved road that runs through beautiful pine forests en route to the ruins of Mixco Viejo, the site of the ancient Poqomám Maya capital. A handful of people work with maguey fibers in the aldeas of Montufar, Sacsuy, and Cerro Alto to make *sembradores* (planting bags), mecapales, *bolsas* (loom-woven bags), alforjas, *enjalmas* (pads), morrales, and *lazo* (rope), all sold by the dozen in Guatemala City's central bus terminal.

My contact in the area was a woman referred to me by a curator at the Museo Ixchel, and Pablo is one of the maguey workers she introduced me to. He lives with his family in an adobe house, about a mile up a dirt road from the main highway. To get there, we walked past adobe and concrete-block houses where I saw men making large wicker baskets and women weaving bright cotton threads on backstrap looms. In front of Pablo's compound is a small store where family members sell produce and soft drinks.

Pablo demonstrated fiber-extraction methods to me, and because water is necessary and it was the dry season, we walked two hours to the processing site. The route began on the main road, turned off down a small path, and eventually became narrower and steeper as we worked our way through pine forests down the mountain to the river. It was a beautiful walk, although, when we climbed back up, loaded down with maguey fibers under the midday sun, I was glad when we reached the main road. Later I learned that Pablo had rented the plants he harvested in 2002, and in 2007 he no longer extracted fibers but bought them from others in town.

Besides showing me fiber extraction, Pablo invited me to learn how to make a sembrador, which his wife, Basilia, and daughter María weave on fixed-vertical looms. These looms are unique to San Juan Sacatepéquez and most likely evolved from the backstrap loom, because the stiff maguey fibers are easier to weave on a fixed loom than on the looser-tensioned backstrap loom. Osborne writes about "bags being made on a small upright loom before which a man squats near the tree from which the loom is suspended," probably referring to this type of loom (1935, 83). Photos document that fixed-vertical looms were in use at least forty years prior to my visit (Osborne 1965, plate 33-c).

The fixed-vertical loom might have evolved from the backstrap loom that O'Neale saw in nearby Rabinal, Baja Verapaz. She writes of men using stick looms to weave maguey bags, but instead of the strap going around their backs, the men sat on low stools with the backstrap under them (O'Neale 1945, 35). This strap position would keep the weaving under constant tension, which is not the case when the strap is around the weaver's back.

I also spent time with the man who made the beautiful alforja in the Museo Ixchel's collection. In addition to saddlebags, he weaves large, square shoulder bags with a triangular top flap, designed and taught to him by his father. He often sells his bags in a small San Juan shop.

While in San Juan Sacatepéquez I heard about maguey processing in the municipalities of Joyabaj and Pachalum, El Quiché, and Cabulco, Baja Verapaz. Deciding to check them out, I went first to Joyabaj where the weekly market is a very crowded, lively affair. As with Nebaj, the traje had changed from bright red to darker, almost brown tones. The most popular morrales in Joyabaj were knitted ones of brown and white wool with vivid geometric patterns and colorful crocheted ones made of cotton. Synthetic or plastic morrales were more common than maguey, and one was made from strips of plastic costales, acquired from one of the development organizations working in the area. There were also white plastic figure-eight-style morrales and redes from Alta Verapaz, mecapales from Concepción Tutuapa, and other maguey products from Santa Lucía la Reforma. Several people told me that the last person who worked with maguey in the area had died twenty-five years before my visit. Even so, a few people mentioned knowing of someone who still processed maguey.

From Joyabaj I took a bus to Pachalum. The route traverses hilly countryside where I saw several people cooking sugarcane in large copper pots over fires. This is the process for making *panela* (low-grade brown sugar). While in Pachalum people mentioned that maguey products from Cabulco and Rabinal are sold at the October 28th fiesta. I also learned that figure-eight-style morrales are made in the vicinity of Mixco Viejo.

In nearby San José de Rodeo, an aldea of Cabulco, Baja Verapaz, market attendees had white plastic morrales, but none made from maguey. I happened to strike up a conversation with a hammock maker from Cabulco, and he invited me to visit his home there. I had read that the people of Cabulco specialized in maguey fiber products between 1550 and 1821 (Feldman 1985, 55) and was curious to see what the current state of the industry was. I took him up on his offer and returned to San Juan Sacatepéquez to pick up my things and catch a bus to Cabulco.

It was a memorable ride. When the bus arrived in the plaza I could see that it was already full, but knowing that few buses traveled this route, and not wanting to go the long way around via Guatemala City, I decided to board. After squeezing into the overstuffed bus I stood for an hour, wedged in between other passengers until a few disembarked and I managed to get part of a seat, where "half a cheek" could fit. (By now my bus-riding muscles were well developed, and this was not difficult.) The unpaved road was horrible. About six miles before we reached the main highway, the bus broke down. The driver and his helper worked on it for about an hour. At this point a pickup truck came by and most of the passengers climbed into the back of it. I waited, primarily because my bags were on top of the bus and there was no choice. About ten minutes later the bus started back up, and we "limped" into Rabinal. From there I took another bus to Cabulco, where Achi is spoken.

I eventually found the hammock maker's house, where he creates *nudo* (knotted), *quadrado* (square), and *tejido* (woven) hammocks. But he wasn't currently working, and so I followed other leads while in the area. One man I spoke with began working with maguey at the age of twelve. Another recounted the fiber-extraction process. Both men said that the plants must be four or five years old before harvesting and approximately eight yards can be leg-spun in a day. I also learned that maguey is cultivated from July through September north of the Río Negro, in an area known as the Embalse (aldeas of Cebollal, Carrizal, Patzijom, Chirramos, Patil, Los Pajales, and Canilla, El Quiché).

Vendors in the Cabulco market said that most of the maguey products came from the departments of Alta Verapaz and Chiquimula. Huipiles and palm petates were more common, made and sold by women in the surrounding aldeas. Petates take two days to make, use about twenty palm leaves, and sell for about twelve quetzales each. Abundant dried palm fronds used for petate making are sold in the market, and maguey fibers were last seen for sale here around 1997.

The last maguey municipality I visited in this region was San José Poaquil, which Osborne mentions as a bag-making place (1965, 176). I planned a day trip from Antigua but initial attempts during Semana Santa did not work out, as buses were already full from the capital and not stopping to pick up passengers in Chimaltenango, where I was connecting. Months later, I finally made it for market day but I did not see any maguey items for sale. However, inquiries suggested that maguey is still cultivated in the countryside.

I admit I felt a bit apprehensive before heading east, for when I lived in Guatemala in the late 1980s, this region had the reputation of being the "Wild East," where macho men with guns ruled the countryside. Fortunately this image had changed, or so I thought, and much later I learned that the East is still considered dangerous, due to drug trafficking (I did wonder about the numerous gun and ammunition stores on the outskirts of Jutiapa). I was glad to be unaware of this at the time, for I met wonderful people and had a fine experience.

5

East and North

I left the central highlands in style. On my way to the capital, the Antigua bus broke down with a horrible noise and much smoke. We were in the middle of Avenida Bolívar, one of the main arterials into Guatemala City, and, as it was rush hour, there was a mass of gridlocked cars, with hundreds of angry, honking drivers. Suddenly, the ayudante's job description included keeping vagrants off of the disabled bus, and he did so with gusto. Not wanting to endanger myself by exiting, I chose to remain and share the fate of the bus with the driver and the few remaining passengers. Since bus drivers are also skilled mechanics, the repairs did not take long, and within an hour we were back in action. As soon as I got off, a first-class bus headed to Jalapa came by, and I quickly stuck out my arm to let the driver know I wanted to board, and he stopped. I hopped on and we were on our way. The route to Jalapa winds down a heavily used highway toward the Río Motagua and then back up again through dry, barren hills.

The Spaniards settled earlier and in larger numbers in the East, primarily because of the lower altitudes, better communications, and indigo, an important cash crop (MacLeod 1973, 230). Thus, acculturation was earlier and greater than elsewhere in the country. Nationally forced labor and taxes or tribute payments in the form of processed cotton and hand-woven cotton cloths were required during colonization, which indicates that looms probably existed in the East, although today backstrap and treadle looms are nonexistent. The absence of these tools is interesting because the "traditional" maguey bag is the bolsa, or loom-woven bag. It's even more curious that in the northern and western regions, where backstrap and treadle loom weaving still occurs (with the exception of the planting bags and saddlebags of San Juan Sacatepéquez), maguey bags are not loom-woven.

The eastern looms are a frame type, which the Spaniards or other traders might have introduced. The designs could be based on "Arawak" or

Amazon looms they saw for hammock making in British Guiana (Guyana) (Drooker 1981, 28–29; Broudy 1979, 98) or the design might have developed independently. Hammocks were woven on "horizontal wood frames" in El Salvador (Osborne 1956, 86), and this could also have been an influence. Frame-type looms are ideal for weaving maguey because it's easier to maintain tension on the stiff plant fibers with a rigid loom.

My first stop was San Pedro Pinula, Jalapa, a Poqomám-speaking municipality sitting in a broad valley at about 3,000 feet. The Poqomám are closely related linguistically to the Poqomchi' and Q'eqchi' of Alta Verapaz and today live in communities scattered throughout the departments of Guatemala and Jalapa. Within the Poqomám Empire the neighboring municipalities of San Pedro Pinula and San Luis Jilotepeque are linguistically classified as Eastern Poqomám Maya and share many of the same traits. Their origin myth takes place on a mountain between the two towns, where God looked down upon them and declared that San Luis would be known for ceramics and San Pedro for palm. This is still true, although both industries are rapidly disappearing (Rodman 2006, 43).

I quickly learned that, besides palm hats, San Pedro Pinula is famous for dairy products. Men lead cows in and out of town daily, and there is a constant battle between those who have cows and those who don't, because of the mess the cows leave behind on the narrow streets. A few older women still wear the traditional cotton blouse and wraparound skirt. Their hair is braided with a narrow ribbon and wrapped around their heads like a *corona* (crown), which signifies that they are married. Women from the aldeas wear *plagadas* (modified dress), a variation of Maya dress worn before the twentieth century. Plagadas are handmade from cheap, imported cloth and most likely evolved because they were less expensive to make or buy than a huipile and corte (Rodman 2006, 48).

Wild maguey grows in the area, but it is not harvested. Instead, maguey fiber from Jocotán is purchased by a couple of families in the aldeas of Agua Zarca and Santo Domingo to make *tenedores* (cruppers) using the technique of ply-split darning. Two brothers, Daniel and José, showed me how to make one. When I visited their home in 2008, José was in the United States and Daniel told me that they stopped making tenedores in 2003, a year after my previous visit.

Maguey fiber is also used as sewing thread for hand-constructed palm hats. In 2002, one man still made them by hand, while everyone else used electric sewing machines (and cotton thread) (fig. 5.1). One of the machine sewers gave me a twenty-minute hat-making demonstration and told me that every three weeks he takes three hundred hats to the central bus

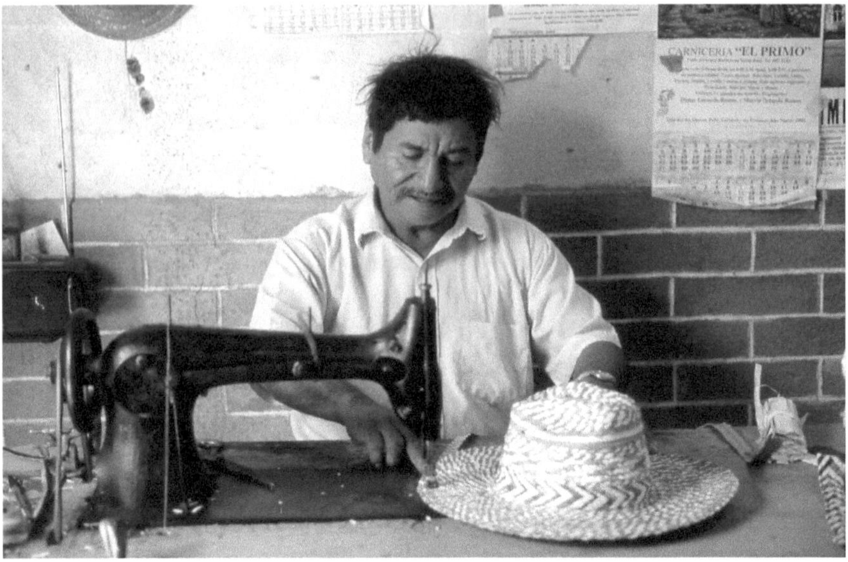

Figure 5.1 Alfredo sewing a palm hat on a sewing machine, San Pedro Pinula, Jalapa.

terminal in Guatemala City to sell. (As a side note to the Pinula hat industry, while I was in Panama I observed a woman using maguey thread to hand-sew plaited palm hats, identical to the Guatemalan style.)

From Pinula I took a bus to Jutiapa, where buses begin their run to Comapa, a town suggested by a man in the Peace Corps office as well as in a 1983 article by Claudia Dary and Aracely Esquivel. While waiting for the bus to depart I wandered through the large daily Jutiapa market and observed many colorful hammocks hanging on the walls. I learned that they came from Comapa. (When I visited the Jutiapa market in 2007, I found that 90 percent of the hammocks were plastic from Jocotán, 5 percent were cloth from El Salvador, and 5 percent were maguey from Jocotán and Comapa.)

The road from Jutiapa to Comapa is paved partway, and buses make the run frequently between Jutiapa and Guatemala City. Between communities the countryside is mostly empty, dry, rolling hills with an occasional wild maguey plant. This was the site of a crime wave in 2005. Many truck drivers were robbed, and an auydante was killed during an attempted robbery the week before I arrived.

I heard about it by accident while eating lunch in a café where I casually mentioned to a policeman seated near me that I had to travel to Jutiapa the following day to meet some people. He sternly told me that I would

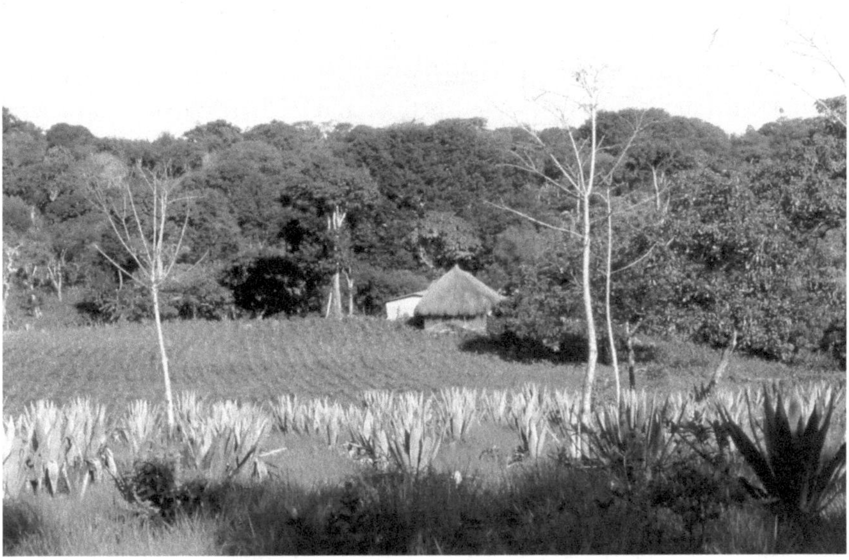

Figure 5.2 Maguey growing in a field near Comapa, Jutiapa.

be better off to stay put because of the little "incidents" taking place on the road. I took my chances and made it safely to Jutiapa, but before returning to Comapa, we convinced the municipal police chief to loan us an officer for our protection. He obliged, and so we had a policeman for the day. In exchange for safe passage, "our man" learned about maguey, something he probably hadn't signed up for when he took the job!

Near Comapa, abundant cultivated maguey plants are seen in front of people's adobe houses in the surrounding aldeas (fig. 5.2). This arrangement is unlike other areas I visited, where maguey plants grow in the distant campo. When I asked why, the reply was that "maguey is considered valuable and there is less chance of theft if the plants are close by." In contrast, twenty years before my visit, maguey was prolific in the surrounding countryside, and there was no danger of theft because everyone had it.

Sitting in the shadow of Volcán Chingo, close to the El Salvador border, Comapa is inhabited by descendents of the Pipil, or *pipiltin*, meaning children or nobles (Fowler 1989, 3). Their ancestors, the Nahua, migrated in waves from Mexico beginning around 900, until about 1350 (Fowler 1989, 49). They settled in pockets along the Pacific coast of Guatemala, El Salvador, Nicaragua, and Costa Rica and spoke Nahuat, an Utno-Aztecan language, closely related to Nahuatl (Fowler 1989, 6). Their capital, now known as Antiguo Cuzcatlán (Antique Land of Jewels), sits just outside present-day San Salvador, El Salvador.

Comapa was known as a "weaver-town" between 1550 and 1750 (Feldman 1985, 59) and as one specializing in maguey-fiber textiles from 1550 to 1821 (Feldman 1985, 55). A few people told me that maguey, along with the knowledge of how to extract and use the fibers, was brought along in the initial migration from Mexico, as maguey use was extensive in their ancestral homeland (Osborne 1956, 13).

The name "Comapa" comes from the word *comali*, meaning *comal* or *olla de agua* (griddle or pot of water). Coffee grows in the area, but maguey is still considered valuable, and at the yearly fiesta the queen holds the title Flower of Maguey, while the runner-up is the Flower of Coffee. One man I spoke with remembered when men and women still wore traje in Comapa. Women's dress was a skirt and blouse made from muslin, which later changed to printed cotton, and they painted their cheeks red. Men wore pants and shirts of the same fabric. He said that the last person in Comapa who spoke Nahuat died in 2003. Today all inhabitants speak Spanish, do not wear traje, and are considered Ladino (a person of mixed blood who speaks Spanish, follows Spanish customs, and wears European clothes). Many told me that those with the "purest Pipil blood" live in the aldeas of San José and Carrizo.

While working in Comapa I stayed with a woman who ran a store that was also a local gathering spot, where I met many people from the aldeas and it was easy to arrange visits with them. One of them was Federico, who demonstrated fiber extraction. He and his father, well-known maguey workers, are the only men in their aldea who still own maguey plants. They sell fibers to the neighbors, who make hammocks, bolsas, *gamarrónes* (halters), and redes.

I also met the husband-and-wife team of Santós and Marco, who shared their knowledge of gamarrón making with me, using the technique of ply-split darning. Many people have horses in Comapa and so horse gear is in high demand.

Maguey fibers are also woven into small bags and saddlebags or double bags on vertical-frame looms, which are found only in this municipality, and I am unsure as to how and when these looms came into existence. In 1945, O'Neale wrote that "bags of this type are separately made between end bars supported to construct small upright looms similar in type to the rug looms of the Near East and the blanket looms of [the] southwestern United States," but she doesn't name the specific location where she saw them, and she did not travel in this region (35, 182).

Hammocks are also made, not on a frame loom, but between two widely spaced poles, which are "planted" in the ground, usually in front of a house.

One of the men who showed me how to make a hammock was Jonas. He was mayor of his community a few years before I met him and lives with his extended family on the outskirts of a Comapa aldea. Besides showing me hammock making, Jonas demonstrated how to start a *red* and, when he began, said that the first loops were "the equivalent of having children."

Another hammock maker I met was Santós, who spent thirty-five years picking cotton on the coast before returning home to make hammocks. In 2001, he made an extremely fine one that won him the *premio a la excelencia* (prize of excellence) in the *primer concurso nacional de artesanías* (first national artist contest). Not only did he gain recognition for his fine work but also prize money. When I returned in 2007, he was still making maguey hammocks and selling them to a Guatemalan living in the United States.

From Comapa I returned to Jutiapa and took a bus to Chiquimula, nicknamed "la Perla del Oriente" (the Pearl of the East) due to its economic importance as an indigo-producing center before the invention of aniline dyes. The name stuck because of its location and status as an important commercial center on a main trade route between the Caribbean and El Salvador. Chiquimula is where buses leave for Jocotán and Olopa, my next sites.

These municipalities are located near the Honduran border, and I visited them based on a conversation with a volunteer in the Peace Corps office, where she showed me bags for sale from Olopa. Jocotán sits at an elevation of 1,167 feet and is locally referred to as "the lowlands," whereas Olopa is at about 5,000 feet and is known as "the highlands" (Wisdom 1940, 8–10; Reina 1969, 120). Abundant palm trees and maguey plants grow in the dry countryside, probably a very different landscape from that at the height of the indigo industry, when Jocotán was a rich center and indigo plants must have covered the hills for miles (Wisdom 1940, 184).

There are two festivals in Jocotán, one for the patron saint Santiago and the other a regional folkloric festival, which was founded in 1986 by Jocotán teachers to highlight Ch'orti' culture (Metz 2006, 197). One of the components of the folkloric fiesta is the election of "Katu Suchij or Reyna o Flor de Maguey" (Queen or Flower of Maguey).

The Ch'orti' are closely related culturally and linguistically to the lowland Maya of Yucatán and Belize (Chol) rather than to the highland Maya of Guatemala (Quiché-Pokam). Scholars agree that the Ch'orti' built the nearby ruins of Quiriguá, Guatemala, and Copán, Honduras, which was their cultural center (Reina 1969, 120). They are self-sufficient with close social and economic ties and are rarely seen outside of the area. One very visible difference from other Guatemalan Maya groups is their traje.

Women's clothing is made from bright commercial fabric and fashioned into full skirts and short blouses with decorative rickrack and ribbons around the neck, armholes, and hem. The dress varies from satin-like blouses and skirts to cotton dresses depending on the aldea a woman is from. In the 1930s, women wore white cotton blouses with "mutton" sleeves and blue skirts adorned with crosshatched lines (Wisdom 1940, 115). Men's traje is made from muslin and sewn with colored threads, which indicate how many children he has had (Metz 2006, 101).

With a long-forgotten weaving tradition, imported fabric has been made into clothing for years (Girard 1949, 263; Wisdom 1940, 64; Metz 2006, 50). A few Ch'orti' women sew their traje by hand, and at one time they had to make their future husbands' traje before they could marry (Girard 1949, 272). Most now buy their outfits from a couple of Ladino seamstresses in town, which was also the case seventy years before my visit (Wisdom 1940, 24). Both men and women wear sandals constructed from leather thongs and tire treads, made by a man in the Chiquimula market. Prior to rubber-tire soles, leather was used, and the Ch'orti' were known as "the man of sandals and the woman of sandals" (Wisdom 1940, 228).

Palm hats, morrales, bolsas, and mecapales are important accessories for men, and the Ch'orti' believe that a mecapal symbolizes God's bearing (or the bearing of a load for a period of time). In the 1930s, Wisdom observed people wearing ornamental or charm necklaces made with strands of maguey and old peso coins, which were believed to ward off evil spirits (Wisdom 1940, 37). He also saw small children carried in maguey net bags by their mothers (Wisdom 1940, 26, 294). I did not see either of these.

Many of Jocotán's inhabitants create maguey, palm, tul (*Cyperus canus*), and *carrizo* (cane; *Montes olyra latifolia* L. and *Arundo donax* L.) items, with products based on the different climate and soil conditions (Wisdom 1940, 22) and a tendency for each aldea to specialize in one item (Wisdom 1940, 23). They sell all over the East, in the department of Peten, and in Guatemala City. The most common palm products are brooms and hats, whereas tul is made into petates and carrizo into baskets. Petate weaving was once considered a royal tradition, with each woven strip representing the reality of time and space, or the Maya cosmovision. Only rulers were allowed to sit on mats, and they were known as the "lords of the mat" (Proyecto Maya Chorti 2002, 1).

Maguey is made into rope, morrales, matates, redes, and hammocks. Many of these items are sold in a few well-stocked Jocotán shops, as well as in the weekly market where middlemen snatch them up. In Jocotán, the people who work with maguey live in the aldeas of Plan de Arada, El Tesoro

Arriba, Canapara Arriba, Canapara Abajo, Tunuco Arriba, Tunuco Abajo, Piedra Parada, and El Potrero. These scattered communities sit on both sides of the steep Río Toroja valley, named for the "bull" sound the river makes during the rainy season. The western side of the valley is known for morrales and hammocks, while the inhabitants of the eastern side produce redes and rope. With steep hillsides and a hot climate, there are few horses in the area and therefore horse gear is not in demand.

With the exception of Concepción Tutuapa, women seem to be more involved in the fiber-extraction process here than in other areas, and one who demonstrated this to me was Ángela. I spotted her one Sunday in the market selling maguey rope with her granddaughter (fig. 5.3). Wanting to take their photo, I approached her to ask for permission. After a lengthy conversation about my project and the product that they were selling, I not only got the okay for the photo but an invitation to visit her home as well.

Ángela is a widow with four daughters and two sons, and they live in a palm-leaf hut without electricity or water in one of the highest compounds of the aldea. When money is needed, Ángela makes and sells maguey rope, and with a fiesta in nearby Esquipulas fast approaching, she was trying to earn some cash.

I also witnessed fiber extraction on the opposite side of the river, with Bonifacia. I met her one day while walking through her aldea with a promoter from Coordinación para el Desarrollo Integral Maya Ch'orti'. Bonifacia is an amazing person, for, in addition to her involvement with many women's groups, she is a midwife and has twelve children, thirty grandchildren, and five great-grandchildren. Bonifacia's knowledge of maguey's history went as far back as her grandfather. When I asked her what she recalled, her face lit up and, with twinkling eyes, she reminisced about her grandfather and his teachings of maguey fiber extraction.

Bonifacia's grandfather gave her a first scraping tool, which was made entirely from wood. It had to be sharpened after scraping every twenty leaves. This tool was probably the same type Wisdom describes, made from oak or quebracho wood, and had grooves cut from the inner portion of the stick (1940, 149–150). Near Bonifacia's house I saw a man scraping leaves, but his scraper, as with all "modern" scrapers, had an old, dull machete blade inserted into a wood handle. He told me that the day before he had scraped 240 leaves, which would provide enough fiber for two hammocks.

Since the region was once known for its indigo, it is not surprising that textile traditions were known, although how and when frame-type looms developed is a mystery. "Vertical upright looms formed by two long poles"

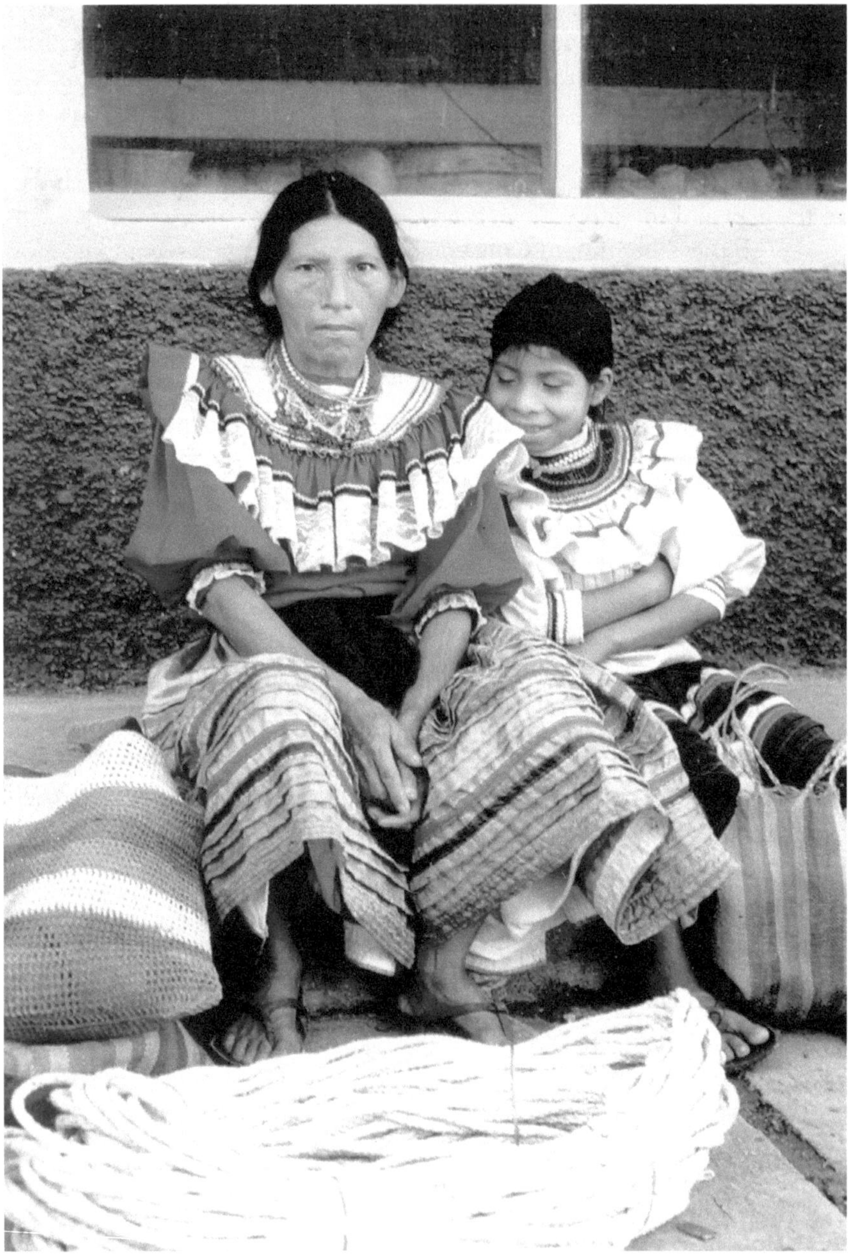

Figure 5.3 Ángela and her granddaughter selling rope in the market, Jocotán, Chiquimula.

were used to weave saddlebags in Jocotán when Osborne was in the area (1965, 179), and Wisdom describes "false looms" for weaving both hammocks and double bags or saddlebags in the 1930s. These looms consisted of four boards held together in a rectangle form that could be brought together or moved apart, depending on the length of the hammock to be made (Wisdom 1940, 159–161), and resembled hammock looms found in South America.

While in Jocotán I observed hammocks, redes, and rope constructed and sold. I did not see any small woven bags made. However, as with Comapa, the hammock "looms" I saw were two poles stuck in the ground, spaced apart to the hammock's length.

In addition, the bag straps Wisdom writes about and illustrates are multiple-strand braiding and possibly ply-split darning (1940, 158–159), also used for mecapales. He states that bag handles are mecapales connected to both sides of the bag. Wisdom does not mention or show diagrams of interlacing, as does O'Neale (1945, 183 with figs. 58g and 58h). They were both in Guatemala during the same time period, but O'Neale did not travel to Jocotán, so either it had not been introduced to the Ch'orti' or Wisdom did not see it.

From Jocotán a pistol-toting bus driver took me to the nearby municipality of Olopa, where I discovered the horizontal loom. This is the only location in the entire country where I saw them, and because they are undocumented elsewhere, except for the mention of a "large horizontal frame used for making hammocks in El Salvador" (Osborne 1965, 178), I am unsure as to when they were introduced. Like the vertical-frame looms of Comapa, they probably exist because stiff maguey fibers are easier to weave on them.

Paula was the first person who demonstrated weaving on this loom and is one of a handful of women creating contemporary maguey products for Asociación de Mujeres Olopenses (AMO). She lives in a single-room wattle-and-daub house with her family, and her loom is located outside the house. It consists of two forked sticks "planted" in the ground.

One of Paula's neighbors had a similar loom set up, except that it was constructed from four separate sticks, instead of two forked sticks, stuck in the ground. On this loom he weaves *bolsas corriente* (quickly made bags), which sell in Chiquimula. He told me two bags used one pound of *pita* (spun maguey fiber), three can be made in a day, and they sell for fifty quetzales a dozen. He also said that weaving is better when the fibers are wet and then proceeded to take a sip of water and spray it from his mouth onto the warp threads as he worked.

Tituque, Agua Blanca, and Tuticopote are the Olopa aldeas where people work with maguey. Tuticopote workers specialize in bolsas corriente, and Tituque is known for bolsas corriente and *bolsas pedida* (bags made to order). The people of Agua Blanca produce mecapales. Horse-related products are not in demand, for there are few horses due to a lack of abundant pastureland, although in the 1940s cintas were made, sold only to Ladinos, as the Ch'orti' did not own horses (Wisdom 1940, 159).

I left Olopa with a sock salesman. He drove by while I was waiting for the bus and gave me a ride to the main highway, where I boarded a Pullman bus headed for the capital. I disembarked at El Rancho, an intersection, where stranded travelers often drink chilled coconut milk while lounging in hammocks to survive the heat. From there I took a microbus to Cobán, the capital of Alta Verapaz. This was my headquarters while I explored the maguey municipalities of San Juan Chamelco, San Cristóbal Verapaz, and Santa Cruz Verapaz.

Here Poqomchíi' and Q'eqchi' are spoken throughout the region, and women's traje includes full skirts made from imported treadle-loom-woven ikat fabric, of the western highlands. The Q'eqchi' don white gauze-like huipiles and the Poqomchíi' wear colorful solid-cloth huipiles, both woven on backstrap looms. Due to the wet climate, traditional houses are made from tree-fern trunks or wood and, today, concrete block is common. I went to Alta Verapaz because I was told about a maguey factory in Santa Cruz, which, I quickly learned, no longer existed.

San Cristóbal and Santa Cruz were both known as maguey-fiber specialist towns between 1550 and 1821 (Feldman 1985, 55) and the traditions persisted. McBryde (1945, 69) wrote that the most important maguey (agave or *Furcraea*) fiber-goods (rope being the most important) supply source to southwest Guatemala is San Cristóbal (followed by San Pedro, San Pablo, and San Juan la Laguna, Sololá, and last by Comitancilla, San Marcos, and Colotenango, Huehuetenango). Osborne agreed and mentioned Alta and Baja Verapaz (especially San Cristóbal Verapaz) as famous for their maguey rope-making industries (1965, 177). In 1990 Alta Verapaz was still considered the most important maguey region in the country (Domingo 1990, appendix p8).

Maguey costales (sacks), used for storing coffee and cacao (chocolate) (Feldman 1985, 56), were made in Concepción Finca. The factory operated between 1910 and 1920 and was located in San Juan Chamelco and Lancing (Finca Magellan). In Santa Cruz, another factory operated from January 1963 to May 1995 and in full swing employed seventy people. This factory produced 20,000 costales and 150 *quintal* (a quintal is one hundred

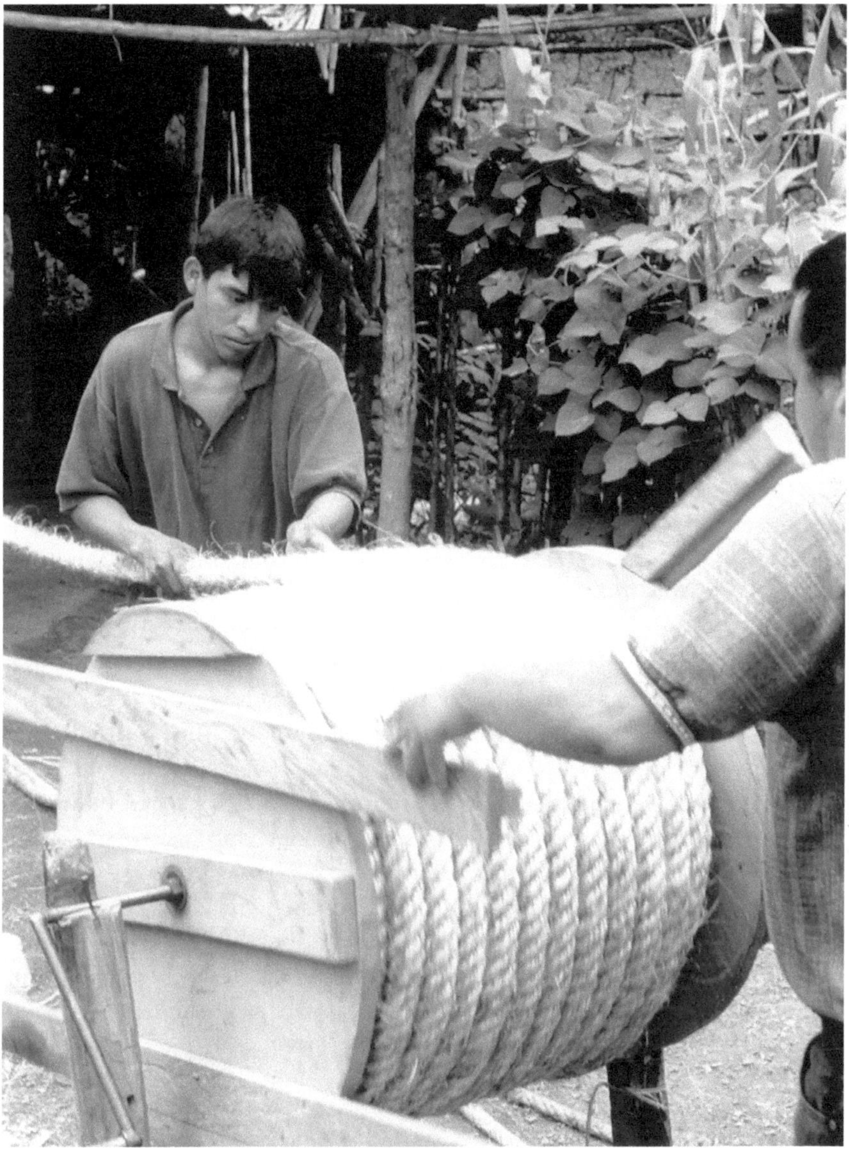

Figure 5.4 César's finished rope, Santa Cruz Verapaz, Alta Verapaz.

pounds) bundles of rope a year, which sold all over Mexico and Central America. When this factory closed in 1995, a few enterprising employees bought some of the equipment and started small operations. César was one of them, and he initiated a small rope-making endeavor in Santa Cruz Verapaz (fig. 5.4).

In 2001, César employed six employees, who were each paid thirty-five quetzales per pound of maguey processed into rope. Five different rope sizes (one inch and one and a quarter, one and a half, two, and three inches) from maguey, silk, cotton, and nylon are made in his factory. Maguey rope sells for more than the other materials and, in the past, maguey fibers were imported from India, Brazil, Mexico, and Costa Rica. In 2002, César used about eighty quintals of local maguey a month. As in other areas, synthetic materials began replacing maguey in the late 1980s, and his customers prefer it because they believe that plastic is biodegradable, which could be because they see it deteriorate in the strong sun. Much of César's maguey rope is sold to the United States and Guatemalan military for creating obstacle courses. Tombstone makers buy the thickest rope and use it for shining the marble headstones.

One of the men César bought fiber from and introduced me to was a sixty-year-old man who makes redes full time. He produces two a day and travels every two months to the Rabinal and Cabulco markets where he sells them for twenty-four quetzales a pair. When I visited César in 2007, he had recently opened an auto repair shop on the edge of town and his brother was running the rope factory.

Another entrepreneur, Alfredo, bought a surplus treadle loom from a *costal* factory when it closed. On my first visit in 2002, the loom sat under only a roof, located in front of his house, making it very easy to find. When I returned in 2007, the loom was less conspicuous, for he had built walls around it. Alfredo weaves "baskets" or round bags, and placemats, but he mentioned that the lack of a steady market is his biggest obstacle.

From Cobán I took a direct bus back to Guatemala City. After many months on the road, I had accumulated several notebooks full of information and diagrams, rolls of undeveloped film, hanks of maguey fibers, balls of spun maguey, maguey products, small tools, and new publications. All of this I transported back to the United States, along with the knowledge of how to make a net bag and other useful items from a plant called maguey.

Part II

The Maguey to Textile Transformation

6

Maguey
The Material

The succulent maguey plant grows in many geographical and climatic conditions and is endemic to a region from the southern United States into South America, as well as the Caribbean. The word *maguey* is an American Indian name (Gentry 1982, 660) used to describe all large-leaf forms of the *Agave,* and *Furcraea* species (Dewey 1965, 3; Camposeco Mateo 1994, 19), in Mexico, Central America, and the Antilles, which is where the name traces back to (Gentry 1982, 23). Both are in the Agavaceae family, of which there are approximately three hundred recognized species, with twenty-three found in Guatemala; of these, seventeen are primarily used (García-Mendoza 2006, 175). *Agave* is the largest genus, with the word originating from the Greek word *agavos*, meaning "noble" or "illustrious."

Other words commonly heard when talking about maguey, especially in the Yucatán, are "sisal" and "henequén," referring to the fiber species of *Agave sisalana* and *Agave fourcroydes* (Camposeco Mateo 1994, 19). Sisal is named for an old shipping port on the peninsula where much of it was exported (Schery 1972, 195). The word *pita* comes from the Caribbean and is spun maguey fibers from Mexico to Brazil (Camposeco Mateo 1994, 19). *Pita floja* is defined as the soft, silky fiber of the *Bromelia* or wild pineapple plant (*Aechmea magdalenae*) (Dewey 1965, 4) and *Bromelia sylvestris* (Squier 1861, 15), or untwisted agave fiber (Osborne 1965, 248). The term *cabuya* or *cabulla* is heard from Costa Rica to Ecuador for some species of *Furcraea* (Camposeco Mateo 1994, 19). The Mayan word for maguey is *sajchi'* (see appendix 4 for spelling variations), which translates to "white" (*sak*), "sun," or "great heat" or "fire" (*quij*), and was related to pre-colonial liquor making (Berger de White and Ivic de Monterroso 2008, 138). The higher-elevation *Furcraea quicheensis* Trel. is called *che'ch* or *chech* (Camposeco Mateo 1994, 20) and *chijute* or *chiyute*. In Concepción Tutuapa, San Marcos, the word *mecate* is heard for *Beschorneria albiflora* Matuda, possibly from the Nahuatl word, *metl*.

Another Nahuatl word, *ichtli* or *ixtle*, is unspun maguey fiber, a common term in Mexico and San Andrés Huista, Huehuetenango. *Pita fina* stands for fibers from *Furcraea gigantean* Vent. *Agave americana, Agave fourcroydes,* and *Agave sisalana,* superior fiber-producing species, all were imported from Mexico into Guatemala by the Spaniards (García-Mendoza 2006, 182). Wild or noncultivated maguey also has many names, including *monte* (San Juan Sacatepéquez), *espalda* (Santa Lucía la Reforma), *cimarrón* (Comapa), and *de sope* (Jocotán).

Another common reference to maguey is *jarcia,* which Greek and Arab sailors used to describe a cord on the mast of sailing ships. The Spaniards adopted the term and brought it to Central America, where it signifies the process of working with maguey (Camposeco Mateo 1994, 31–34). Maguey earned the name "century plant" when it was transplanted as an ornamental to the cool climate of northern Europe and the flowering stage took much longer than in its warmer, native region (Standley and Steyermark 1952, 105).

Early inhabitants in the maguey region discovered a multitude of uses for the plant, and some, such as the Otomi of central Mexico, relied entirely on maguey to survive (Granberg 1970, 12). Maguey was potentially responsible for the foundation of agriculture in the region (Gentry 1982, 4).

Some of its products include food and drink in many forms. The roasted white "meat" in the lower stem and base of the leaves can be eaten, as well as the boiled flowers (Gentry 1982 6–7; Camposeco Mateo 1994, 26; García-Mendoza 2006, 183). Honey or syrup drunk fresh in a drink called *aguamiel* (honey water) is processed from the "heart" of the plant and was one of the few sources of sugar in pre-colonial times (McBryde 1945, 142). Fermented sap becomes pulque and is distilled for mescal and tequila (Gentry 1982, 8; Valenzuela-Zapata and Nabhan 2003). The alcoholic effects from pulque were incorporated into religious and mystical ceremonies, which, to the Aztec, were so important that several gods were associated with the rituals. The most famous was Mayahuel, the goddess of pulque (Gentry 1982, 8–9). Some Maya women made a tea with maguey roots, quina bark, rosemary leaves, and nutmeg to aid in expelling the placenta during childbirth (Wisdom 1940, 288). The sap from sisal contains hecogenin, a source material for the synthesis of cortisone (Schery 1972, 195). Vinegar, syrup, fish poison, soap, and adherent for whitewash, also made from the inner juices, are more by-products of this amazing plant (Gentry 1982, 21; García-Mendoza 2006, 184).

Shelters, fences, ladders, and aqueducts can be constructed from the strong stalks and leaves, which may also be burned for fuel or cut up and made into paper. Maguey paper was valued so much for its toughness and durability that hieroglyphic records and ritual calendars were painted on it in ancient Mexico, and much later, an 1830 law required that maguey paper be used for recording laws and legal documents, including passports (Squier 1861, 4, 35). In San Pablo la Laguna the thick maguey trunks are cut into sections and made into drums (Berger de White and Ivic de Monterroso 2008, 176–179), as they were in other areas (Irigoyen 1950, 17). The leaves provide material for fertilizer, livestock fodder, and medicine (García-Mendoza 2006, 184) and can also be used as a carving canvas, often lasting forever, as Vicente Fernández's song "La Ley del Monte" describes (2000). The sharp spike found at the tip of the leaf is fashioned into needles.

Another use is *cercos vivos* (live fences), where maguey is planted in rows along the contour of a hillside and prevents erosion by trapping dirt that would otherwise wash down the steep slopes during heavy rains. Cercos vivos can also contain livestock. Besides being planted as a fence, it is often grown as an ornamental.

Finally, the part of the plant providing one of the greatest uses regionwide is its fiber, found deep inside the thick, tough leaf. Indigenous people from the southwestern United States throughout Mexico, Central and South America, and the Caribbean made cordage, nets, bags, baskets, mats, blankets, clothing, and sandals from it for centuries. The Aztec goddess of henequén and cotton was Tlazolteotl-Toci (Johnson 1971, 311).

A plant's transformation into woven artifacts is a worldwide occurrence leading to regional specialization in many unique environments. The knowledge and processes in the "land of maguey" probably developed over many years. Archaeological finds and written evidence indicate that some of what I saw has been practiced for generations, whereas other aspects evolved over time as politics, economics, and needs changed. As to the initial discovery that usable fibers existed inside the thick maguey leaf, there are a few possibilities.

In one Maya legend, Zamna, the highest priest of the Itzaes, was out looking for medicinal plants when he scraped his hand on an agave spine. In anger he cut the leaf that hurt him and discovered that deep within the flesh were long, white fibers. He later proclaimed: "La vida nace en compañía del dolor, origen de todo bien en la tierra. A través de una herida se nos ha revelado la existencia de una planta que será de gran utilidad a mi pueblo, a los hombres y las mujeres del mundo maya [Life begins in the company of pain, the origin of everything good on the land. Because of a wound, the existence of a plant with great uses for my town and the men and women of the Maya world was revealed]" (Benítez 1962, 55; Irigoyen 1950, 8–9).

Others believe that maguey fibers were first revealed when a bolt of lightning struck a leaf, searing away the green fleshy outer skin and exposing the white inner fibers. In yet another tale, maguey was "discovered" by survivors of a landslide who were living along the shore of Lake Atitlán. After heavy rains they relocated as the water level rose, and when the lake receded, they returned. In the exact location of their huts, maguey was growing, and the population soon learned that many things could be made from this plant. Ever since, it has been cultivated and considered very important, even sacred (Berger de White and Ivic de Monterroso 2008, 136, 139).

Unfortunately, material evidence of ancient maguey textiles is rare, because most examples disintegrated due to the destructive effects of a tropical climate on vegetable fibers. However, some glimpses of early maguey use include oblong stones, which were used for fiber extraction and have been found in archaeological sites around Guatemala (Osborne 1965, 174), and maguey textile fragments discovered in caves (Johnson 1971, 302–310); evidence from the El Riego Phase of Tehuacán, Mexico, shows that as early as 6500 to 4800 BCE maguey textiles existed (Ventura 1996, 11; Schery 1972, 167).

Archaeological excavations also uncovered maguey use in the Zapotitán valley of what is now El Salvador, where lava from Loma Caldera's 600 CE eruption preserved maguey gardens and spun fibers in the city of Joya de Cerén (Sheets 2002, 187). Other clues come from *The Annals of the Cakchiquel*, a description of this Maya group's early life, where maguey fibers are mentioned (Recinos and Goetz 1953, 68). More evidence comes from the Middle American codices, ancient manuscripts that detail pre-Hispanic attire of the Aztecs, Tlaxcalans, Tarascans, Mixtecs, Borgia, and lowland Mayas, including many maguey items (Anawalt 1981, 27, 30, 33, 100, 137).

Maguey plants are monocotyledons, and the fibers are classified as leaf or foliaceous (hard), which are larger and stiffer than bast or cortical (soft) fibers, such as flax, nettles (*Urtica*), hemp (*Hibiscus cannabinus* or kenaf, ambari), and jute (*Corchorus*) from dicotyledonous plants (Dewey 1965, 1). Maguey is also known as an "inside grower" because the inner-fiber vascular bundles provide an important structural component in the leaves and carry food and water from one part to the other (Mauersberger 1947, 369; Squier 1861, 23; Schery 1972, 171). Other leaf fibers are abaca and esparto and those from the pineapple plant (Emery 1966, 5).

Growing from the middle outward, a maguey plant, in the form of a rosette, takes approximately two or three years to mature, at which time the leaves lay in a horizontal position (Schery 1972, 195). It is a monocarpic plant, meaning that it flowers once in a lifetime and then dies.

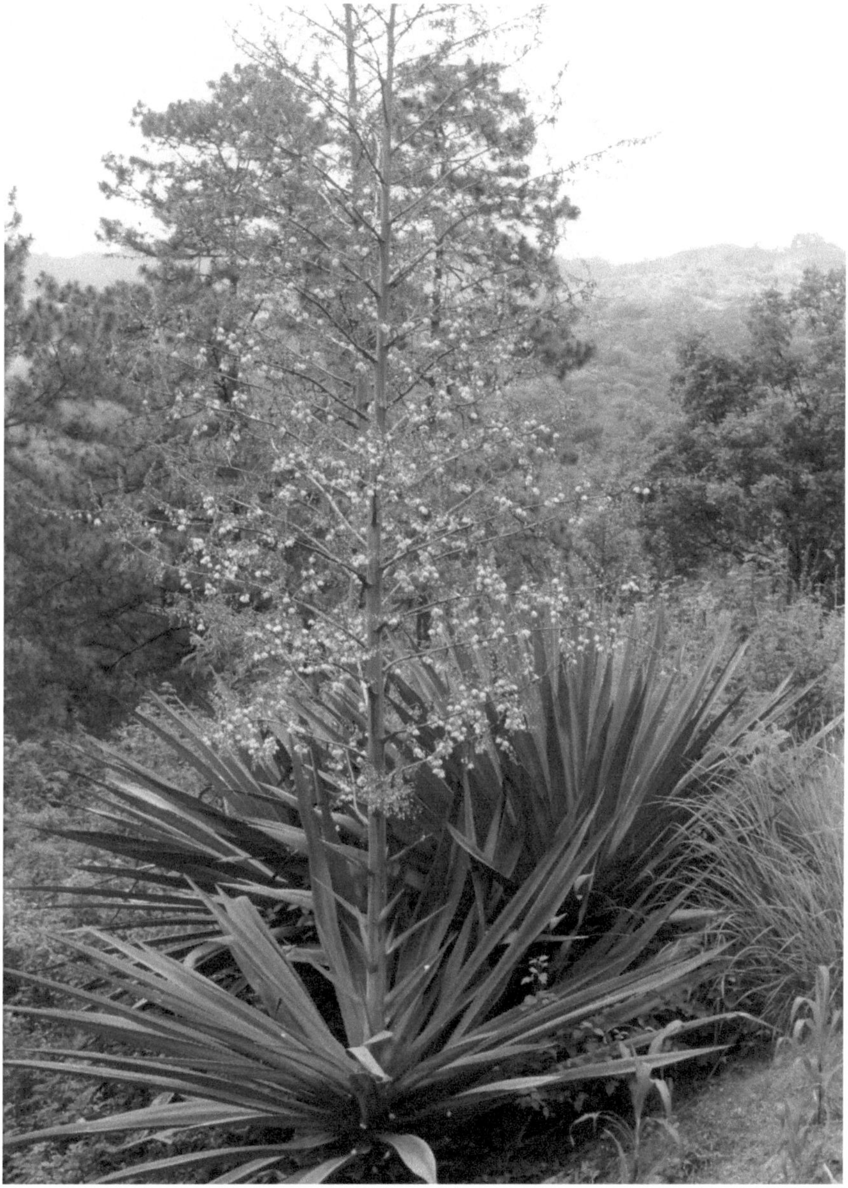

Figure 6.1 Maguey plant (*Furcraea* sp. nov.) in flower, between Jocotán and Olopa, Chiquimula.

When the flowering stage begins, a thick stalk emerges from the plant's center (fig 6.1). It grows up to a foot a day and can reach heights of thirty feet (Standley and Steyermark 1952, 106). If not harvested, some species will flower in ten to twelve years, whereas the same species will live for up to twenty-five years if the leaves are cut. As explained to me, the longer life span occurs because, when the leaves are cut, the plant's "energy" goes into growing new leaves instead of producing a flower.

Reproduction can occur sexually and asexually. For sexual reproduction the pollinators are often bats and hummingbirds (Gentry 1982, 42); asexual reproduction occurs when the "mother" plant sends out rhizomes, which form bulbils or small shoots (hijos), also sometimes called "suckers." The species sending out abundant shoots are preferred to those growing from seeds for they retain genetic material (Gentry 1982, 300), and superior plants can be selected for propagation. In many locations, such as San Pablo la Laguna, new shoots are transplanted about a foot apart, when the moon is full (Berger de White and Ivic de Monterroso 2008,141). The moon phase is also important to the Otomi for planting (Salinas 2000, 87). Thus begins a new cycle for the maguey plant.

7

Fiber Extraction

The first step in transforming a thick, green leaf or *penca* into a beautiful textile product is to extract the inner fibers. Ancient agave fragments discovered in a Mexican Coxcatlán cave show evidence of chewed-away soft tissue to extract the fibers (Schery 1972, 168). This method was probably very tedious and not very pleasant. Today, the leaves are cut at a 45-degree angle in a complete circle close to the stalk, beginning from the outside of the plant and working inward (photo essay fig. 1). This allows the new inner ones to "fall" into place, which occurs after about a month. Cutting the leaves at an angle also prevents water from collecting in the plant's "heart" during the rainy season and rotting the interior. Generally, each plant grows about three hundred leaves in its lifetime (Schery 1972, 195). Fiber yield is about 4 or 5 percent of the green leaf weight (Schery 1972), and a standard harvest of approximately thirty-five leaves per plant equals about one pound of dry fiber.

Depending on the species, one leaf can be up to five feet long and five pounds in weight. Before the process of extracting the inner fibers begins, the sharp end points, side teeth, and thick head are trimmed off, usually done near the plant (fig. 7.1). The knife or machete must be sharp for this, and in the American Southwest, special igneous and metamorphic rock "agave knives" were used (Parsons and Parsons 1990, 301). In contrast, the scraping tool should be dull, or the fibers might be accidentally cut while removing the nonfibrous material.

Nationwide, most people believe that the extracted fibers are stronger if leaves are cut around the time of the full moon, for if they are cut too soon the fibers will be tough, and if cut when the moon is small, they will be weak. The timing of the cutting could have a scientific basis as, right around this time, there is more water or sap in the leaves. This applies to the harvest of other crops as well, such as rushes, and "if they are not

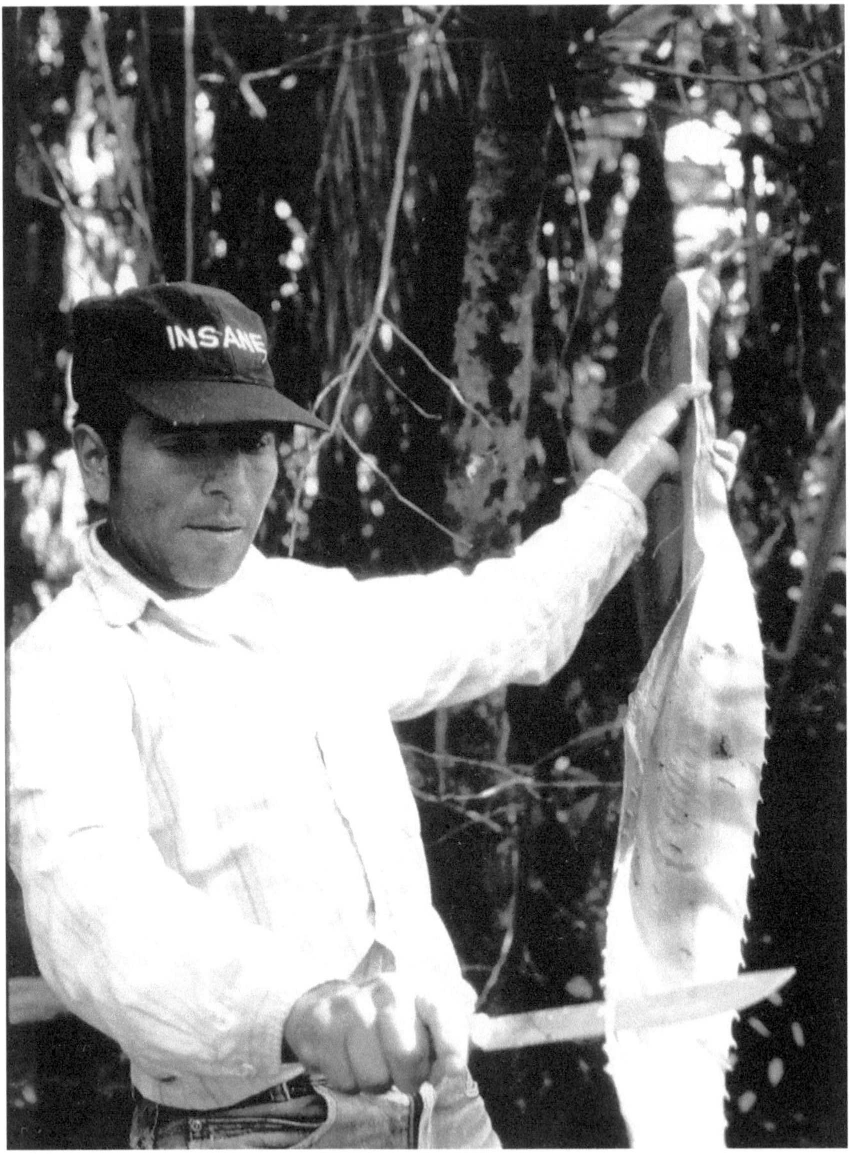

Figure 7.1 Nicolás trimming a maguey leaf, San Juan Cotzal, El Quiché.

cut when the moon is full, the mat will fall apart and the maker's reputation will be ruined" (Kelsey and Osborne 1961, 71). Panamanian *chácara* (string bag) workers also harvest maguey leaves during the full moon (De Luco 2001, 64).

Fiber characteristics depend on the plant's age, bundle position, species, and growing conditions (Schery 1972, 195). Young plants have more pliable leaves and are easier to work with, and the inner leaves have the finest fibers (Parsons and Parsons 1990, 316). *Furcraea* is known to have softer fibers than *Agave* (García-Mendoza 2006, 183).

For the best-quality fiber, extraction should be done as soon as possible after cutting and spun within a year of processing while the fibers are still soft and not brittle. I was told that cut leaves can sit a few days but, if they are left much longer, they will dry out and be unusable. Squier discovered that within twenty-four hours of harvesting, leaf sap begins to ferment, turns to acid, and starts to weaken the fibers. Also, if left in the sun for a few days, the sap dries and stains the fibers (Squier 1861, 20). Even so, one maguey harvester had a pile of leaves cut four months prior that he said could be used for making rope.

One hazard of working with maguey is the caustic sap or juice that leaks out of a cut leaf when scraping. The liquid can cause acute contact dermatitis, which is very painful. Another fiber-extraction hardship occurs in areas where water is required, which often means long walks to a river or lake. In addition, the acid in the scraped flesh poisons fish, and when leaves are soaked in ponds, the oxygen is removed from the water, also killing the fish (Edholm and Wilder 1995, 4). Where roasting or steaming is part of the process, firewood is needed but is often difficult to find or expensive to buy, and could contribute to deforestation.

Because of the difficult and dangerous work, in most locations men harvest the leaves and extract the fibers. Nationwide, the methods vary depending on species, climate, intended fiber use, and custom. Where wild maguey grows, it is used only for emergency string as needed in the field.

After the leaves are harvested, the remaining plant consists of a bulging white body with a few green spiky leaves pointing upward from the middle, resembling a large pineapple (fig. 7.2). Parsons and Parsons suggest that the fiber-extraction methods by which leaves are softened or decomposed prior to scraping are older because bone, wood, or stone tools would be less effective for removing the hard, raw maguey leaf flesh than the metal scraping tools developed later (1990, 298). They also state that extraction sites were usually located near maguey plantations due to the enormous amount of unneeded bulky, heavy waste products from the fleshy leaf and spike parts; transporting the harvested dry fibers away from the extraction sites is much easier than transporting the whole leaf (Parsons and Parsons 1990, 297). While this is true in many locations, not all follow this model.

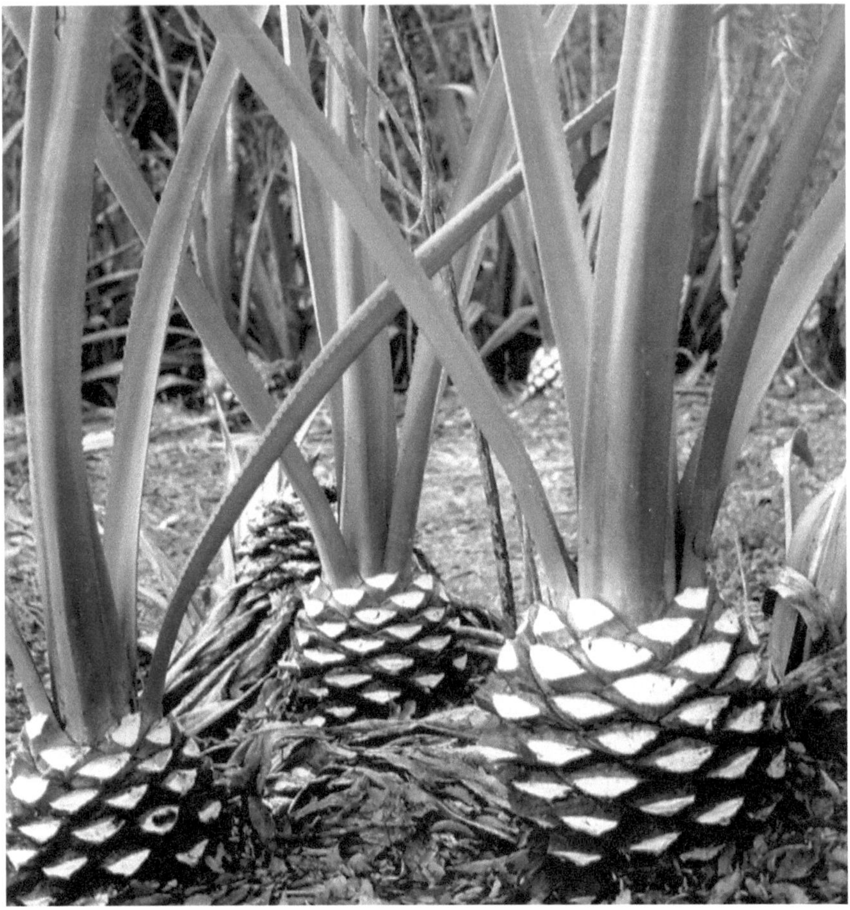

Figure 7.2 Harvested maguey plants.

The following are methods by which maguey fibers are extracted from the leaves in Guatemala.

Retting (Fiber Extraction with Water)

Retting is a common fiber-extraction process of soaking leaves in standing or slow-moving water, which rots or softens the tough exterior so that it can be easily removed. The soaking duration depends on the weather, water temperature, and maguey species. If the leaves are left too long in the water, the fibers will be weak, and if they are undersoaked, the flesh may be difficult to remove.

Lake Atitlán provides ample water for soaking leaves near San Pablo and San Marcos la Laguna. Because the maguey plants grow on the steep hillsides around these communities, after the plant is cut and trimmed the leaves and tools are carried to the fiber-extraction areas, near the lake and at a small year-round stream on the outskirts of town. Beside the stream bank is a small dirt path, and scattered along the bank are large stones. There are also a few stones near the lake. These are what the maguey leaves are pounded on, to shred the tough outer skin and pulpy substance, making it easier to remove. By opening up the skin, the water penetrates inside the leaf more readily, and the outer skin comes off more easily if the pieces are smaller (photo essay fig. 2). Leaf remnants are found next to stones where recent processing has occurred, but when the stones have not been used for years, they sit alone in silence.

The beating stones are considered private property and, in one case, an elderly man was worried that his would be stolen, because none of his sons or grandsons planned to continue the family maguey tradition. He went to the mayor for advice and was advised to paint his name on the stone. Still worried, he had me take his photo next to the stone as further proof. When I visited San Pablo in 2006, all of the pounding stones were gone, washed away by the torrential rains of Hurricane Stan in 2005. By 2007, a few were back in place, but not nearly as many as before.

Stones have not always been used, and when Lothrop visited San Pablo la Laguna in the 1920s he noticed that the leaves were beaten with short, wooden mallets against a block of wood (1929, 121). Some forty years later, when Osborne saw the process, the blocks of wood had been replaced with stones: "Even though quantities of grooved stone bark beaters have been unearthed in various parts of the country, only well-worn river stones are used for beating" (1965, 174). Osborne also observed a large forklike wooden tool being used to remove the pulp from the fibers (1965, 174). In addition, Lorthop saw men separating fibers by careful combing (1929, 122). In 2001, I did not see forklike wooden tools or combing of fibers in this area.

Today, the six-pound *machucador* (hardwood mallet or *pilón*; Osborne 1965, 174) leaf pounder is made from *guachipilín* (*Diphysa robinioides* Benth. [Fabaceae family]). In San Pablo, after pounding, the shredded leaves are placed in standing water for a minimum of four hours. These small man-made ponds are in the natural streambed; where the stones are close to the lakeshore, leaves are soaked in the lake. O'Neale observed rock-outlined basins built into the lakeshore where leaves were soaked for several days (1945, 20).

Once the soaking is complete, the soft maguey leaves are placed, one at a time, on top of a wooden plank sitting directly on the ground. The pulpy leaf exterior is scraped away with a long-handled, dull, metal-bladed tool. The remaining fibers are dried overnight and then put into bundles, rinsed, and dried again the next day. When Lothrop visited in the 1920s, he observed that the San Pablo men took the bundles of scraped fibers out into the lake with their canoes to wash and "remove all traces of stickiness," perhaps to avoid killing fish living close to shore (1929, 121–22). I did not observe this practice, which probably means that less maguey is being processed and/or there are fewer fish.

A father-and-son team, who work full time with maguey, processed thirty-three leaves in two and a half hours. Another man, who began working with maguey when he was ten, said he could process forty leaves in four hours. Yet another said he usually cuts thirty-five leaves from each plant every three to four months.

In San Marcos, a softer fiber is desired and so the extraction process is slightly different than that of San Pablo. The contrast is that the shredded leaves are weighted down with rocks in the lake and left for four or five days rather than a few hours. This allows the outer pulp to rot further, making it easier to remove, and creates a softer fiber. Soaking longer also removes more of the caustic sap. After soaking, scraping is the same as in San Pablo, on a plank with a long-handled, dull, metal scraper (fig. 7.3). The fibers are washed and dried in the sun. Sixty years prior, McBryde observed that leaves were soaked in the lake for two weeks (1945, 69).

In San Juan Sacatepéquez, fiber extraction also requires water. Plants are not cut until they are five years old and are harvested only three more times during their life span. The best fibers come from eight-year-old plants. Growing among pine and oak trees, the cut and trimmed leaves are taken to the river where manmade ponds have been constructed in the riverbed.

One day when I observed the extraction process, there was a pile of harvested leaves on the riverbank when we arrived. They had been cut, trimmed, and stripped by hand two weeks earlier. There were also bunches of twenty-six *manojos* (handfuls of leaves, with the exact number dependent on the maguey species; here each manojo has approximately seven leaves) near a small manmade pond. These leaves had been removed two days before our arrival. Another bundle of twenty-seven manojos was in the pond, representing a day's worth of cutting and trimming. The soaked bunches were twisted together in compact bundles of about seven soaked leaves and then pounded on a rock with an oak mallet (fig. 7.4). They were not scraped. The fibers were then rinsed and dried in the sun.

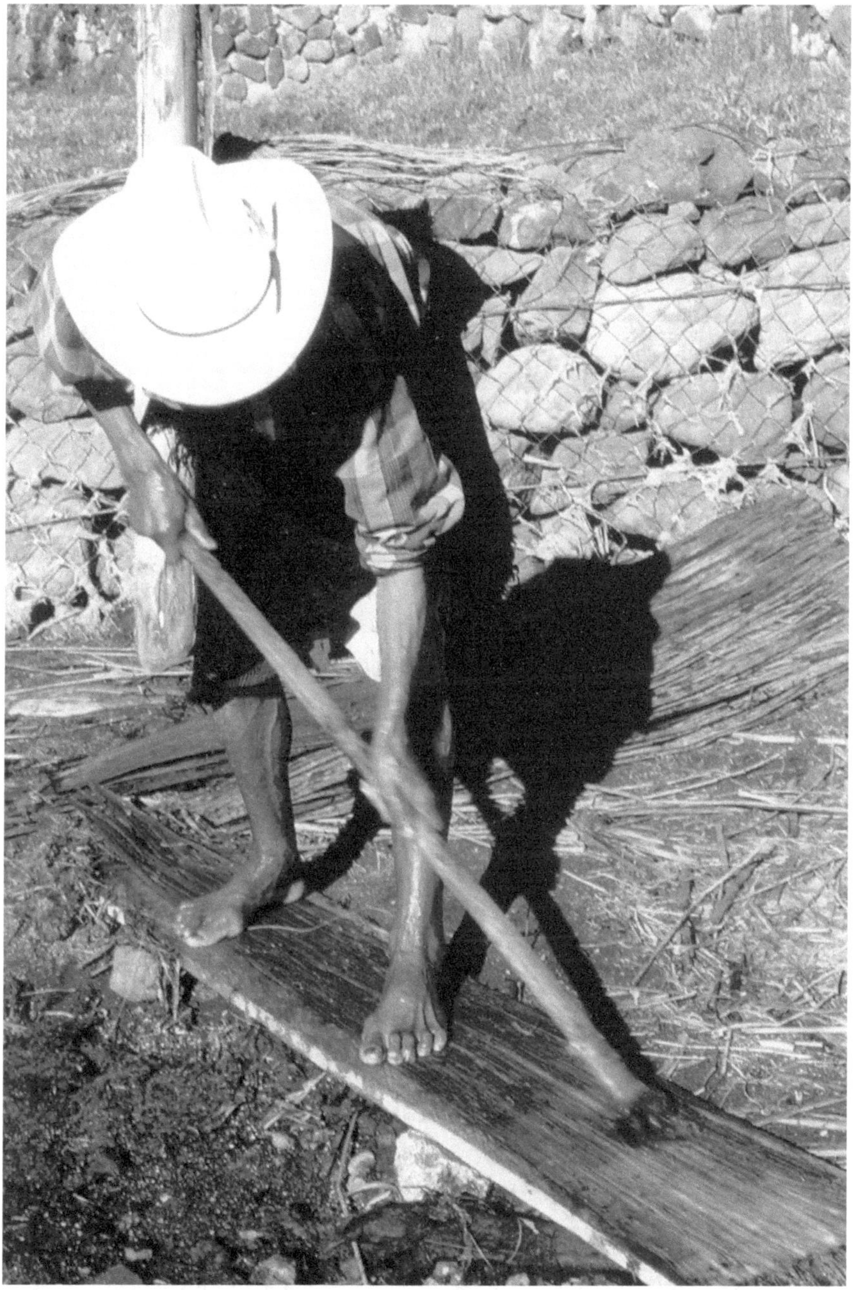

Figure 7.3 Lucas scraping a maguey leaf, San Marcos la Laguna, Sololá.

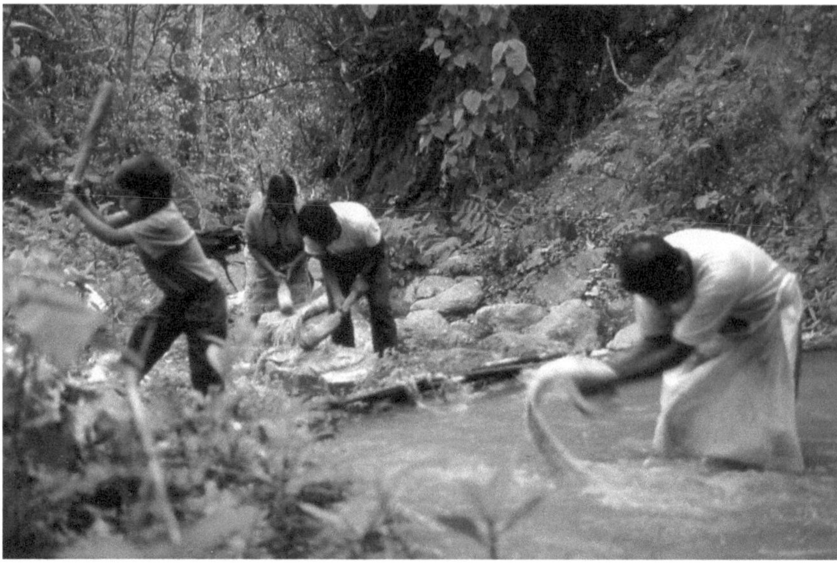

Figure 7.4 Pablo and family processing maguey leaves, San Juan Sacatepéquez, Guatemala.

San Andrés Huista is another location where water is essential for fiber extraction. The maguey fields are near ponds, which are close to a stream. The pond that I saw was not in the streambed, as elsewhere, but was a manmade ten-foot-diameter pit filled with water diverted from the nearby stream.

The cut and trimmed leaves are held down with a stick, under the worker's foot, and then cut into one-inch strips with different cow-bone tools, depending on whether the bottom or top end of the leaf is to be sliced (fig. 7.5). After many leaves are split, they are put in the pond and stones are placed over the leaves to hold them down. The leaves are left for two to three weeks (longer if the weather is cooler) in standing water. After they are soaked, the leaves are washed in the river, pounded on a rock, and dried. Like San Juan Sacatepéquez, they are not scraped.

As in San Pablo and San Marcos la Laguna, the ponds and pounding rocks are considered private property and can be rented out to others. The first harvest comes after the plant is four years old and must be right after the full moon, but the timing of subsequent cuts does not matter. Each plant can be harvested three times a year, for twenty years.

Soaking in water was also mentioned as part of the process in Cabulco, where the leaves are pounded, put in water for eight days, and then scraped on a vertical board. In the Embalse area, maguey leaves are cut and

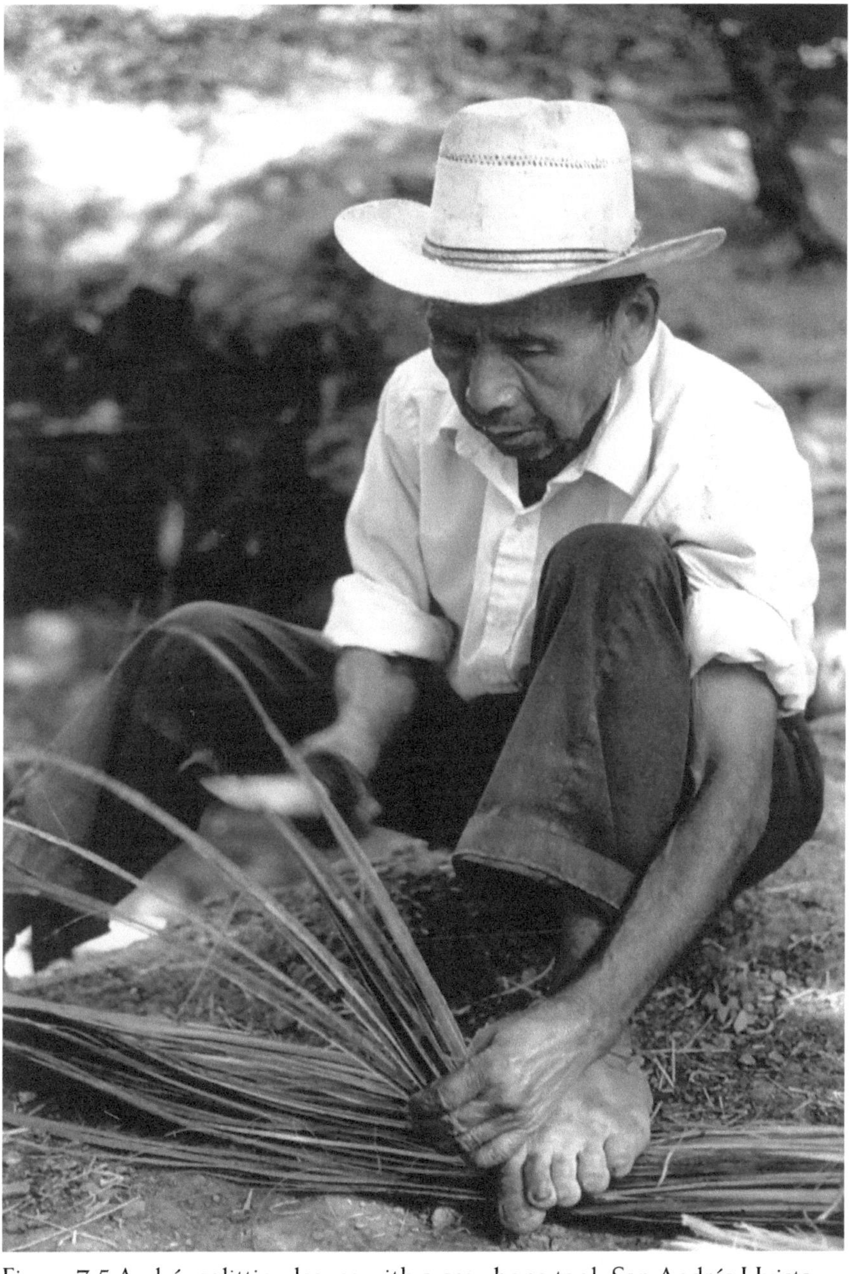

Figure 7.5 Andrés splitting leaves with a cow-bone tool, San Andrés Huista, Jacaltenango, Huehuetenango.

put in water for three to four days, pounded, and then scraped or pulled between two sticks; see Acatenango below, under the "Raw (Fiber Extraction without Fire or Water)" subhead. Pulling between two sticks (or rebar) was also mentioned in other nearby areas as an added step to make the fibers softer.

Roasting (Fiber Extraction with Fire)

Roasting over a fire, combined with burial, also rots the maguey leaf exterior, enabling easy fiber extraction. As with retting, the length of burial time depends on the weather and species. Roasting should be done within a week of harvesting the leaves, while they are still green, or they might burn (Parsons and Parsons 1990, 298).

Roasting is the common method in San Juan Cotzal. I was told that three-year-old plants are considered the best because their leaves are wider and thicker. Usually, after harvesting, a plant is not cut again for a year. The trimmed leaves are put on top of a strong, green-wood frame, above a hot fire, and constantly turned as they roast (fig. 7.6). The leaves are "ready" when they turn black and the inner juices begin leaking out. After removal from the fire, the leaves are buried under mulch in a shady spot, where they sit for about two weeks, which is also when the pile begins to smell. Periodically adding water and turning the pile every so often can speed up the process.

The rotted, soft individual leaves are placed on a horizontal wooden plank and scraped with a *cabo* (long stick with a dull metal blade at the end). The pointy end of individual leaves is scraped first. Then groups of five to ten leaves are tied around a stick, held down with the worker's foot, and the thick ends are scraped. The remaining fibers are washed in rainwater and, finally, dried in the sun. I was told a good scraper is able to process six pounds of fiber in one day.

Roasting is also the custom in San Juan Chamelco, Alta Verapaz, and Santa Lucía la Reforma, Totonicapán. In contrast to the San Juan Cotzal method, a vertical board is used for scraping in San Juan Chamelco, and in Santa Lucía la Reforma, after the leaves are singed they are buried in a three-foot-deep pit of mud. The leaves are left to rot for fifteen to twenty days, depending on the weather (longer if cooler). After the leaves are removed from the pit, the rotted pulp is scraped away from the fiber with a solid wood scraper against a horizontal board. The fibers could be softer than those of San Juan Cotzal because of the difference in processing or the

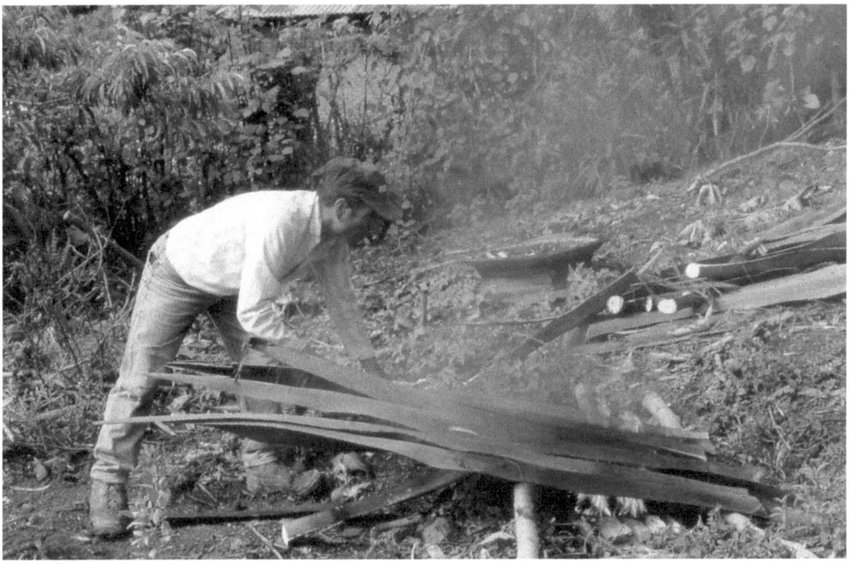

Figure 7.6 Nicolás roasting maguey leaves over a fire, San Juan Cotzal, El Quiché.

different maguey species. A Chamelco man told me that, of the local fibers, the whitest and cleanest come from Tucuru, the next best from San Pedro Carchá, and finally the least desirable from San Juan Chamelco. Fibers extracted from roasted leaves were found to be finer than those obtained using other methods (Parsons and Parsons 1990, 317).

An additional step I witnessed in Alta Verapaz was the sorting of fibers into three classes with a "comb" or hackler (a block of wood with nails extending outward). Often, as the fibers are pulled through the comb, *sebo de res* (beef fat) is added, making the fibers soft and able to adhere together better. Sometimes a mixture of diesel and oil is applied, which brightens the fibers and makes them insect resistant.

In Olopa, the higher-elevation mescal is harvested, which has longer, thicker leaves and produces a softer fiber than the maguey verde and blanco found in lower Jocotán. Working with mescal doesn't sting as much as maguey verde and the fibers are easier to spin because they are softer. The leaves are usually cut two days after the full moon and the same plant is harvested once a year.

The fiber-extraction process involves pounding the head of the cut and trimmed leaves, roasting them over a fire, and letting the leaves rot for five days. Two or three leaves are scraped at a time on a vertical board, then washed and dried in the sun. Putting leaves in water for five days and scraping was also mentioned for fiber extraction and is the preferred method, because the fibers come out whiter than those extracted by fire. But when the streams were full of fish, this method was discouraged. One man had one thousand leaves rotting in a field. He said it took him a day to cut and trim them and a day to roast them, and it would take him three days to scrape, which he does using two or three leaves at a time.

Steaming (Fiber Extraction with Fire and Water)

In Concepción Tutuapa, mecate grows throughout the countryside in straight lines between fields of corn or wheat. The plants are harvested at all times of the year, and during any phase of the moon, when they are one to seven years old. A few men said that other maguey species grow in the area but no one ever uses them. I did not actually see other varieties.

The leaves grow up to four feet long, and to harvest them, they are pulled from the plant by hand. In this way the new leaves will "fall" into place. If the leaves are cut with a machete they won't grow back correctly, because the entire leaf must be removed, and this is easier to accomplish by pulling rather than cutting. After the leaves are removed, the pointed tip and thick "head" are cut off with a machete. This species has thin, pliable leaves, no side teeth, and little caustic sap, making the job much easier than in other locations, which could be one reason women are the primary maguey workers here. The trimmed leaves are bundled into groups of three, folded into thirds, and placed in a clay pot with water. A fire is built under the pot and the leaves are steamed for four hours and then left overnight in the pot to cool.

After removing the leaves from the pot (fig. 7.7), they are arranged into bunches of about ten leaves and each bundle is tied with string to a vertical board, usually leaning against the outside of a house. The maguey worker scrapes the fibers in a downward motion with an old, dull machete blade, holding it on both sides and being careful not to break the long fibers during the scraping (photo essay fig. 3). This removes the green outer skin, leaving behind the inner white fibers. The remaining fibers are washed in rainwater and dried in the sun.

Another municipality where fibers are extracted from steamed leaves is Chichicastenango. Like Concepción Tutuapa, the maguey species harvested has pliable leaves and no side teeth. The difference is that, after steaming, the soft pulp is scraped away by a person standing behind, rather than in front of, an upright board and scraping down. This is the same method that

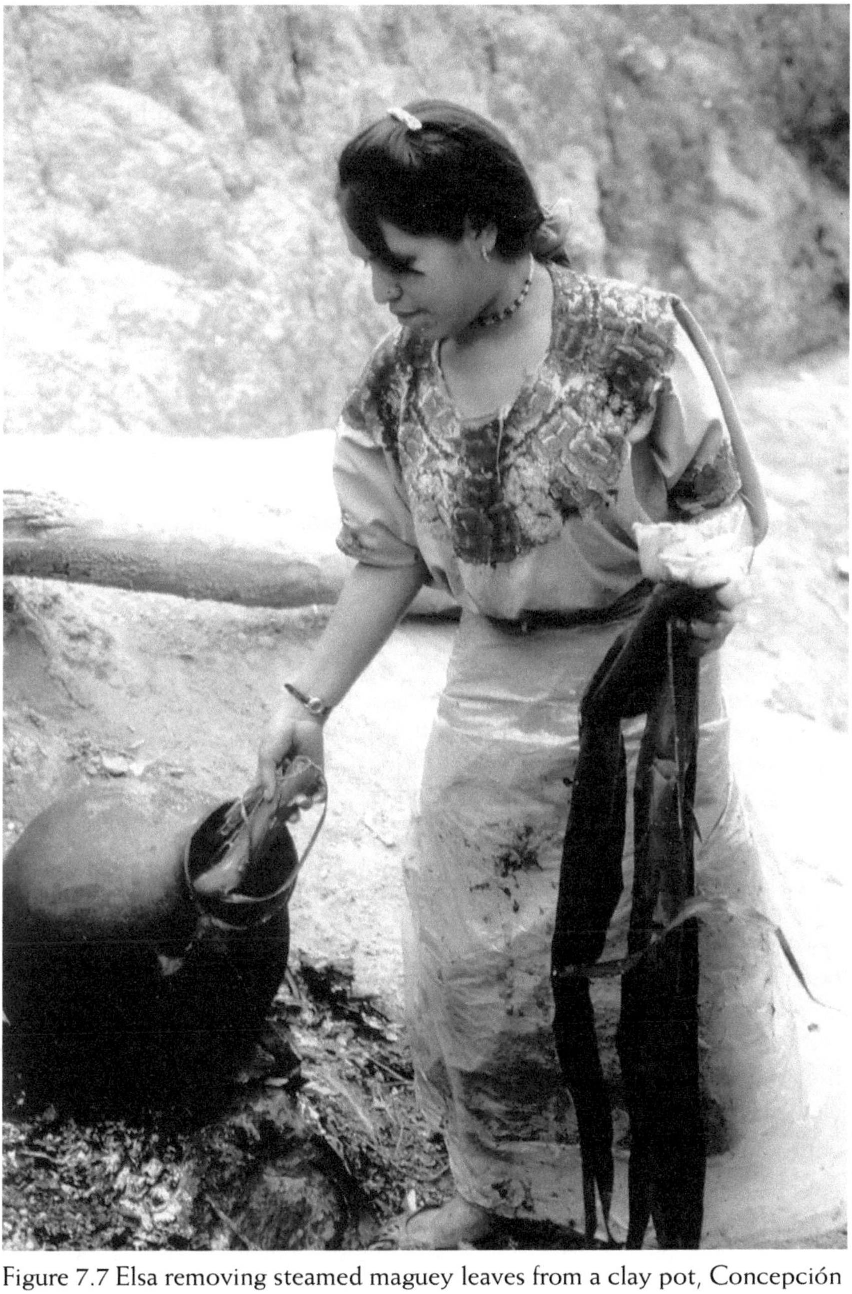

Figure 7.7 Elsa removing steamed maguey leaves from a clay pot, Concepción Tutuapa, San Marcos.

I saw in Chiapas, Mexico; it is also documented in central Mexico (Parsons and Parsons 1990, 226). The fibers are then washed and dried in the sun. In contrast to Concepción Tutuapa, men are the maguey workers in these locations. O'Neale learned that some Chichi men preferred hizote (*Yucca elephantipes* Regal) fibers to maguey ones because they felt that, when prepared correctly, they were finer and stronger (1945, 22).

Raw (Fiber Extraction without Fire or Water)

Comapa is one location where raw leaves are processed. Of the two species harvested, maguey verde is tall and has green leaves with finer, softer fibers, but there is more acid in the leaves; maguey blanco is shorter, with greenish-blue leaves and whiter fibers. Maguey blanco is considered first rate, whereas maguey verde is known as second class. In addition, maguey blanco sends out more shoots from the "mother" plant than does maguey verde, which is desirable. The same plant is harvested every three months, after the leaves drop, for ten years.

After the point, head, and side teeth are trimmed from the cut leaves, a wooden spike is usually inserted at the thick end of the leaf (fig. 7.8). A rope goes around both ends of the spike, and the leaf is placed, pointed end down, onto a vertical board leaning against a tree. The leaf is scraped with short motions from the bottom upward using a tool called a *mania de palo*, which has a curved wooden handle and an old machete blade inserted in a long depression. One man told me that he scrapes about six leaves an hour, or one and a half to two pounds of fiber a day. After a few leaves are scraped, they are put in bundles of three and the "heads" are pounded with a wooden mallet against a small stone. Then the bundle is retied to the board and scraped again in a downward direction to remove the remaining pulp from the thick ends (fig. 7.9).

Due to the hot climate, scraping is only done between five and seven in the morning, because the caustic sap enters the skin's pores more readily when the worker is sweating. The early-morning schedule is very strictly adhered to, and when I tried to get anyone to demonstrate the extraction method outside of this time frame it was absolutely impossible.

Once scraped, the leaves are dried in the sun. It is believed that water rots the fibers and turns them yellow, so they are never washed, although a quick rinse probably won't damage the fibers as much as soaking them for extended periods of time, as Squier discovered (1861, 12–13). I was told

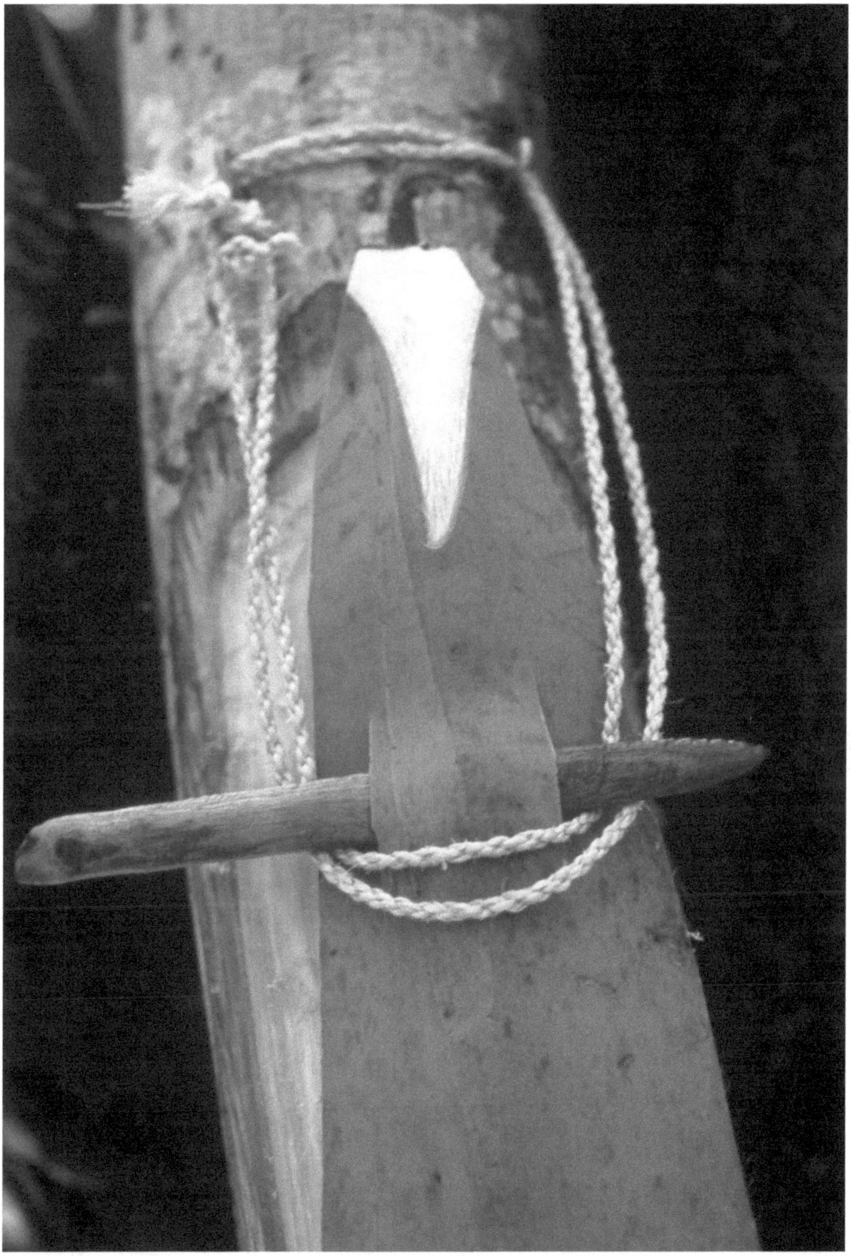

Figure 7.8 Spike through a maguey leaf, Comapa, Jutiapa.

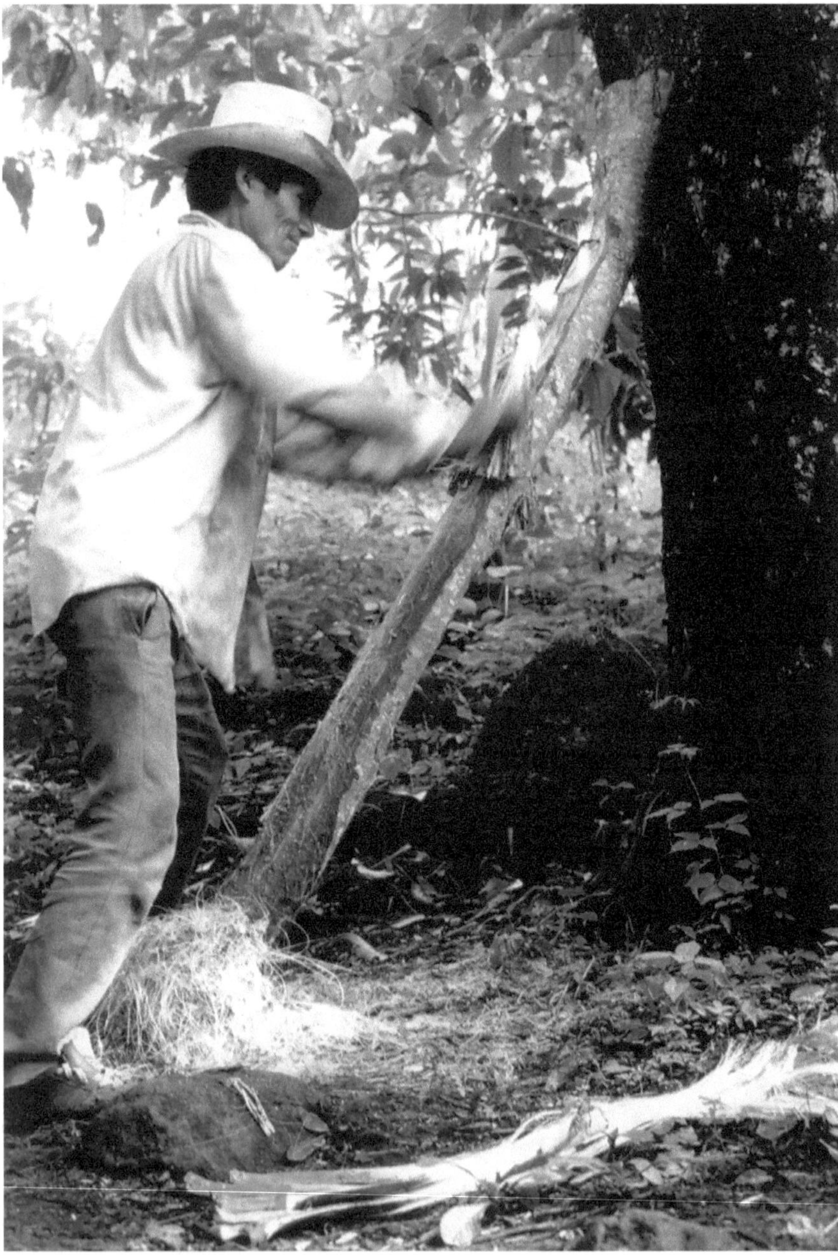

Figure 7.9 Federico scraping raw maguey leaves, Comapa, Jutiapa.

that the softest fiber comes from leaves harvested two to three days after the full moon.

Jocotán is another place where fibers are extracted from raw leaves. Here *maguey de castillo* (verde) and *maguey de seda* (blanco) are harvested. The maguey de castillo has rougher fibers, and so it is used mostly for making rope and hammocks. I also saw *maguey sin espina* or *maguey extranjero* (sisal), which was planted about forty-five years prior to my visit, as part of an economic development project, and while this species has softer fiber, no side teeth, and doesn't sting much, there are few plants left because most were stolen over the years. In the 1930s, each household had a maguey orchard with between fifteen and twenty plants (Wisdom 1940, 64). By the time I visited, beginning in 2001, probably half of the households still had maguey gardens.

In contrast to Comapa, both men and women harvest and scrape leaves at all times of the day, even though it can get quite hot in Jocotán. The scraping tools resemble the type that I saw in Comapa. Another difference is that the "head" is not pounded on a stone with a small wooden mallet, but instead the leaf is "thrown" over the processor's shoulder onto a stone (fig. 7.10). The results are the same, breaking down the hard thick "head" of the leaves. Next the leaves are tied to a vertical board with a piece of string and scraped, point end first. A wooden spike is not used. The maguey de seda stings more than the others and is harder to scrape but has longer fibers (about two and a half feet long). The extracted fibers are not washed, just dried in the sun for four days (photo essay fig. 4). Wisdom noted that if this is not done, the fibers won't hold their twist and will straighten out after spinning (1940, 153).

In Acatenango, maguey fibers are also extracted from the raw leaves. Here, as elsewhere, the leaves are cut and trimmed with a sharp knife. Next the leaves are thrown against a pole to break down the hard outer skin, which mirrors the procedure in other locations where solid objects, such as stones, are used. Then a sharp knife is used to split the leaves into one-and-a-half-inch strips. Next the individual strips of five leaves bunched together are pulled between two sticks or two pieces of rebar, held tight as the leaves come through (fig. 7.11). The rebar or sticks are tied to a pole, which is tied at an angle to a small tree.

I never heard a specific term for this process, but Osborne (1965, 175) called it a "*parenquima,* a Spanish word adopted by the Indians for the soft part of the leaf, as well as for the device that separates fiber from pulp. Parenchyma is the fundamental tissue which makes up the bulk of fruit pulp or pith of stems." Osborne also mentions another kind of parenquima

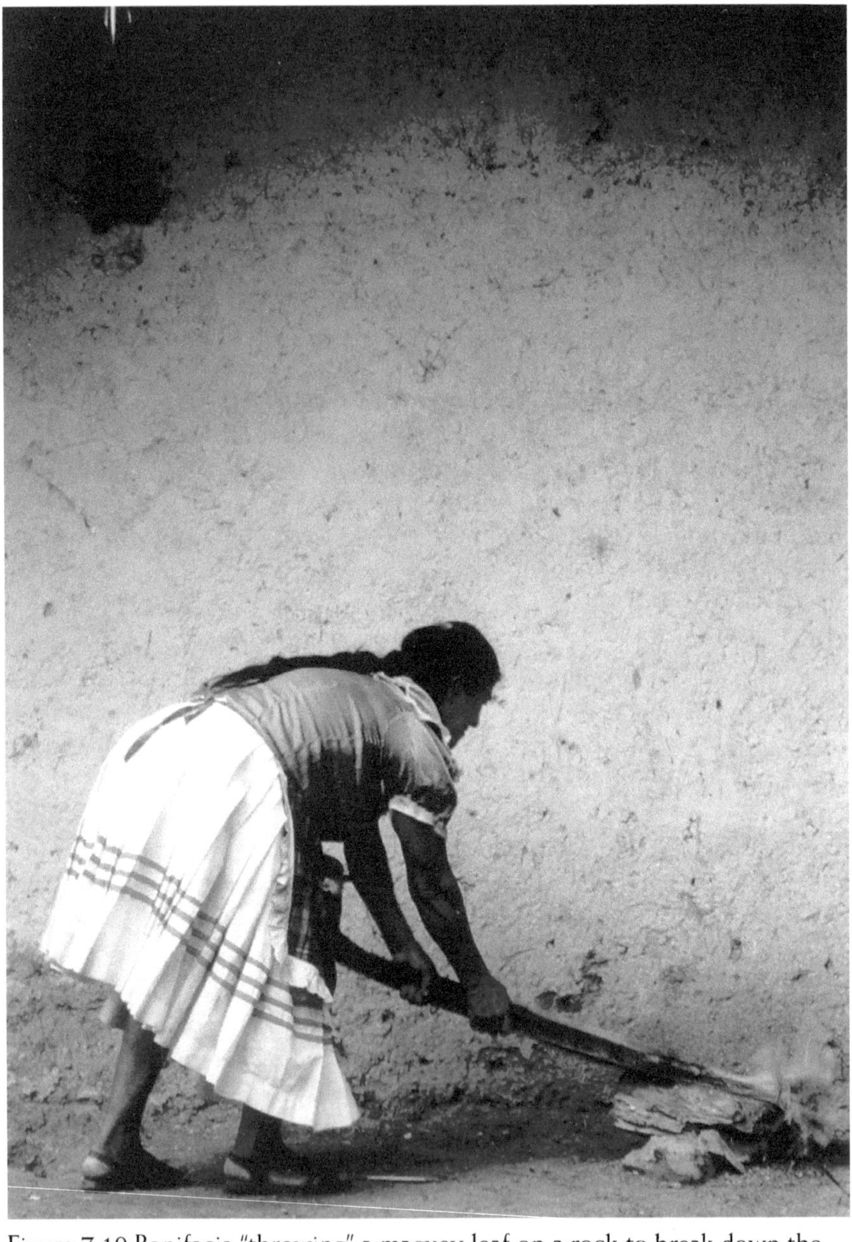

Figure 7.10 Bonifacia "throwing" a maguey leaf on a rock to break down the thick outer skin, Jocotán, Chiquimula.

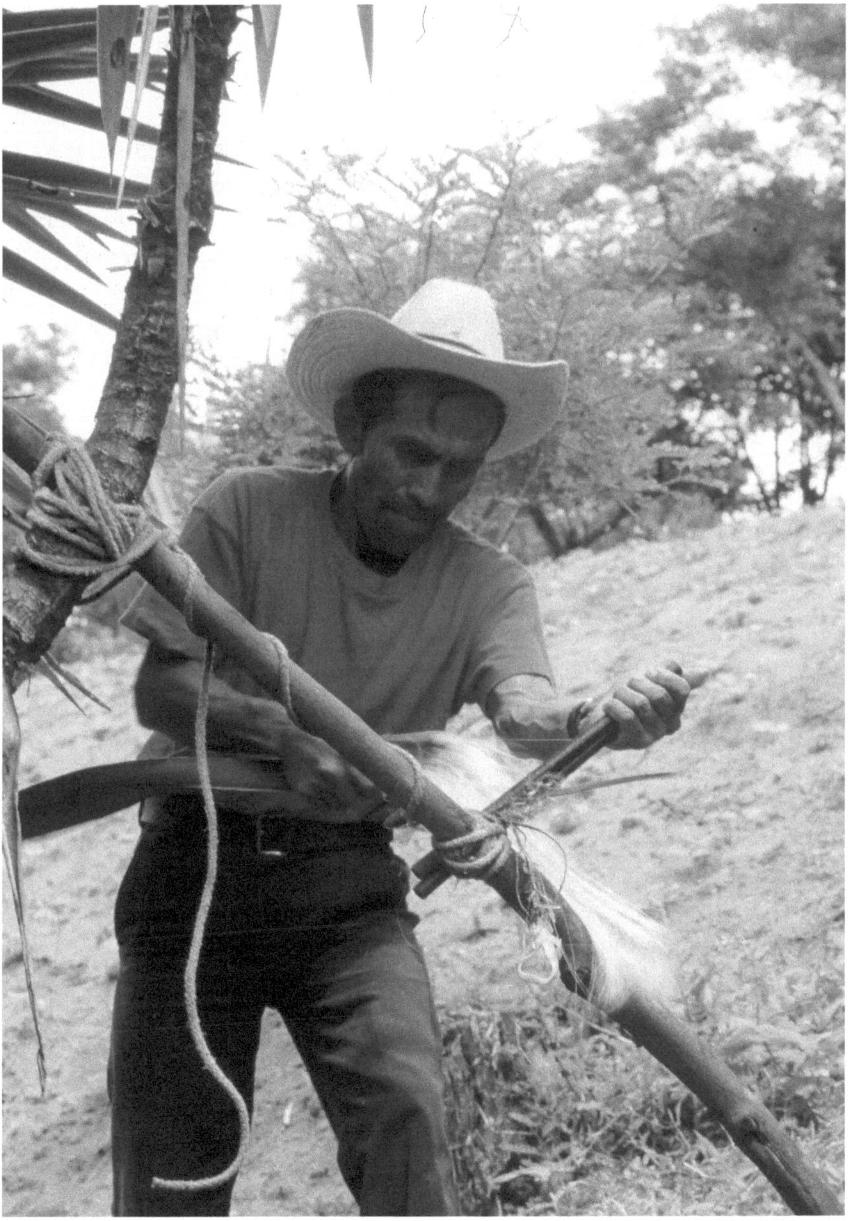

Figure 7.11 Eugenio pulling maguey leaves between two pieces of rebar, Acatenango, Chimaltenango.

made of two small, rough, unfinished boards where leaves are passed back and forth to separate the fibers, which is similar to what I saw in Pachalum; there, the fibers are extracted from the leaf by pulling them between two rough sticks, which the workers call a *sacador*.

The friction from the rebar or sticks removes the outer pulp, leaving behind the white fibers. The fibers are not washed, as it is thought that water will turn the fibers black. The maguey workers also believe that allowing the fibers to rot will weaken them (Osborne 1965, 175). Squier found both of these suppositions to be true, as he noted that "water stains and weakens the fibers" (1861, 12–13). After extraction, the fibers are dried for about four hours in the sun. Here I was told the best time to harvest maguey is in the fourth quarter of the moon, as the fibers will break if cut at other times. Each plant can be cut every two to three months for twenty years.

I found the same fiber-extraction method in Joyabaj, El Quiché, where leaf strips are pulled through two closely placed sticks or rebar. Sometimes, to soften the maguey leaves they are soaked in water for eight days prior to stripping. Fiber extraction by pulling between two parallel sticks was also documented in Jacaltenango (Anderson 1978, 44; Camposeco Mateo 1994, 32).

In central Mexico, extracted maguey fibers were bleached, which was accomplished by soaking them in soapy water for two or three days (Parsons and Parsons 1990, 179) or in a solution of water, commercial soap, and lime juice for a day (Parsons and Parsons 1990, 174). After either process, the fibers were dried in the sun. I never heard of this being done in Guatemala, although white fibers are preferred over yellow ones.

In all cases, after the fibers are extracted they are ready for use. In some areas, unspun fibers are in demand and are sold in regional markets. In other places, trade occurs between neighbors, and many who own and process fibers use them in their own work. For those who want color, the next step is to dye the unspun fibers.

8

Colors and Patterns

White or light yellow is the natural color of maguey fibers; for any other color, dyes are necessary. In most of the country, bright pink, green, yellow, and purple aniline dyes are favored because they are easy to use and relatively inexpensive, plus the colors are bright, which is preferable. The powdered dye is sold by the ounce in local stores. However, aniline-dyed maguey fibers will run if they get wet, because a mordant (fixative) is not used. I never saw unspun or spun aniline-dyed maguey fibers for sale in a market or shop.

Before the invention of synthetic dyes by Sir William Henry Perkins in 1857 (Osborne 1965, 35), natural materials were gathered and processed to color maguey fibers. Concepción Tutuapa is one municipality where I saw a natural dye commonly used with maguey. Red-brown is the color manufactured and it comes from the cambium, or inner bark, of the *aliso* (alder) tree (*Alnus acuminate* [Betulacae family]). To get the strongest color, first the right tree must be found. Incisions are cut into the bark of several trees, and after a few minutes, a deep red color appears. The tree with the darkest shade is selected; chunks cut from this tree are chopped into smaller bits and placed in a pot of water along with cut-up avocado pits.

The mixture is boiled for approximately forty minutes. Then the maguey fibers are added and "cooked" for about an hour. The pot is removed from the fire and liquid *cal* (powdered hydrogenated limestone combined with water) is added to the mixture. (Cal is very important in Guatemala and is sold in every market, as it is used for softening corn kernels when making *masa*, or cornmeal. Raw cal is mined and fired in high-temperature kilns before it is usable.) Sometimes ashes are added. Wisdom noticed that cal softened the fibers and ash altered the color (1940, 154). The fibers are left overnight to cool.

The alder dyes do not run because the tannins in the bark and the avocado pits are natural mordants (although I discovered later that sometimes the dyes are not completely fixed, most likely because the dye bath did not reach the proper temperature, the time spent in the dye bath was too short to set them, or the excess dyes were not rinsed out entirely). To best prevent colors from fading or running, when the clean, damp fibers are placed in the bath, the water temperature should build slowly to a temperature of 190°F (88°C) and be held for about an hour, during which the fibers should be gently rotated, to even out the exposure and give better coverage (Wipplinger 1996, 50).

Unspun aliso-dyed fibers are sold in the local markets alongside bags with the distinct red-brown stripes. O'Neale wrote of bags with red-brown stripes from the Cobán area (1945, 182), and I would guess that aliso was used to obtain this color.

In 2002, I became aware of a small resurgence of naturally dyed maguey in Jocotán and Olopa, where dye workshops sponsored by Capacitador de Artesanos and AMO took place. The knowledge of extracting colors from *palo de pito* (*Erythrina* sp.), coconut (*Cocus nucifera* L.), *nance* (*Byronima crassifolia*), guava (*Psidium guajaba* L.), and *cedro* (*Cedrela mexicana* M. Roem.) plants was reintroduced. Interestingly, indigo or añil (*Indigofera guatemalensis*), so important to Jocotán prior to synthetic dyes, is now imported from Momostenango because it no longer grows in the area. By 2009 all of the products sold by AMO were made with naturally dyed maguey.

Wisdom also observed natural-dye use in Jocotán during the 1940s. The most important dyes were yellow from the bark of the logwood or *palo de campeche* (*Haematoxylum campechianum* L.); red from *palo de Brazil* (*Haematoxylum brasiletto* Karsten); and yellow from the roots, leaves, buds, twigs, and stalks of fustic and coral trees.

The design on most bags is stripes, with stripe direction determined by the construction technique. Looped bags always have horizontal stripes, because they are worked in spirals and the colored strands follow the working line. Most bags have three to six stripes that are three or four rows wide, although I have seen bags with up to ten stripes. Woven bags have vertical stripes because the woven panel that forms the bag is always warp faced. Woven bags usually have two to four str ipes. Linked bags also have vertical stripes, although when constructed the stripes are horizontal.

Sometimes powdered dyes are mixed with water to form a thick paste, and accent colors or village scenes are painted on items. In 2005 a Jocotán vendor had several loom-woven bags with painted floral and animal designs on both sides. He said they were a "rediscovered" stash from San

Pedro Sacatepéquez, Guatemala. In 2004, I visited a man in San Pedro who still paints woven maguey bags, as well as weaves maguey rugs. His products are sometimes found at Nim Pot, a shop in Antigua.

Another color-adding method observed by Lothrop (1929, 125) and Osborne (1965, 174) in San Pablo la Laguna seems to have disappeared. They both describe a method for dyeing fibers by stretching the strings between two sticks and rubbing them with color. This produced parallel lines of alternating colors, resembling tie-dye. I did not see or hear of this process and, in 2007, I showed photos from Lothrop's article to many people in San Pablo without finding anyone who remembered it. After the desired fibers are dyed, the next step is to spin them.

9

Spinning Fibers into Strands

Spinning joins together short fibers into continuous usable lengths, with the simplest strand called a single or single ply (Emery 1966, 9). These can be plied by twisting (or "laying," a rope-making term) together two or more singles, with the direction opposite to that of the single (Emery 1966, 10). Maguey strands are 2-S or clockwise-twisted single strands plied together to form a 2-ply Z, or counterclockwise twist. The angle of twist can vary, and the more twists per unit of linear measurement, the tighter the strand (Emery 1966, 11).

Maguey fibers were most likely first spun on the leg or between a spinner's hands. These simple techniques require no tools and produce a strong, tightly spun strand of any desired length or thickness. Spinning is a very important step for, if the strand is not tight and uniform, the finished product will be weak, inflexible, and not aesthetically pleasing.

Leg Spinning

Leg spinning is found nationwide and is the common method in Concepción Tutuapa, due to the nature of the bags made in the region, where large amounts of fiber cannot be spun in advance because the end of the working strand must pass through a very small space. Usually only about three yards are spun at a time. Leg spinning was very tedious and time consuming for me to learn, with each yard of spun maguey fiber taking me about three hours to spin.

The women make spinning look very easy and they get beautiful, even strands every time. The challenge is having the right amount of fiber, along with the correct motion and tension of the hands. First, the spinning motion occurs, with the left index finger and thumb holding tight the joined end, and the flattened palm of the right hand moving down the leg, pushing two separate parallel groupings of fibers together, in a clockwise direction. Then, quickly the left hand releases the strand, and the two strands are joined in a counterclockwise "twist" by moving the right hand back up the leg. A second, quick spin up the leg tightens and strengthens the strand. Sometimes the free ends tend to ply loosely together after spinning and must be separated before continuing.

One of the most difficult steps is joining, or splicing in new strands, and if this is not done correctly the strand will be weak. The secret is to add small amounts of fiber frequently (Edholm and Wilder 1995, 3). To attain a "rhythm," many women line up several singles before plying them together. It is also important to spin the head (thick end) into the tail (thin end) of the previous strand.

To aid with spinning, small cloth bags filled with dried pine pitch chunks are "hit" on the leg, before the placement of maguey. This action emits a fine, sticky dust that helps the fibers adhere better. Some spinners prefer wet pine pitch, saliva, or ashes of burnt termite tailings for the same effect.

Because maguey fibers are rough and spinning on bare skin can cause bleeding, I liked to joke that red blood was just another natural dye. Others also wrote about the rough fiber irritating the skin and causing bleeding (Osborne 1965, 177; Lothrop 1929, 122). To protect the bare skin, yet give the same friction, many spinners place a piece of leather, tire, inner tube, or shoe bottom on their upper leg.

In Concepción Tutuapa, where women are the primary spinners, an interesting spinning accessory is found in most households: the *palo* (stick), a three-legged, five-inch-wide or so length of wood with a piece of leather, smooth piece of tire, inner tube, or old shoe bottom attached to the top. It serves as a "leg" by setting it on the ground next to the spinner or on their lap. I believe that this tool was developed to make spinning maguey easier for women of this region, because they wear a corte and it would be hard for them to spin on their upper thigh. I do not know when the palo came into existence, and there is evidence of women using the bare skin of their lower leg in the 1980s (Rodríguez 1985, 63). I did not see this tool anywhere else in the country (fig. 9.1).

I reserved a palo from an older woman but had to leave it until a future trip because it was an awkward shape to transport on the crowded buses. When I returned a couple of years later and asked how much it was and to whom I should pay, there was a bit of confusion. It seemed that the daughter of the palo owner had loaned her mother a chicken to be returned upon successful propagation. The palo was used as collateral since they

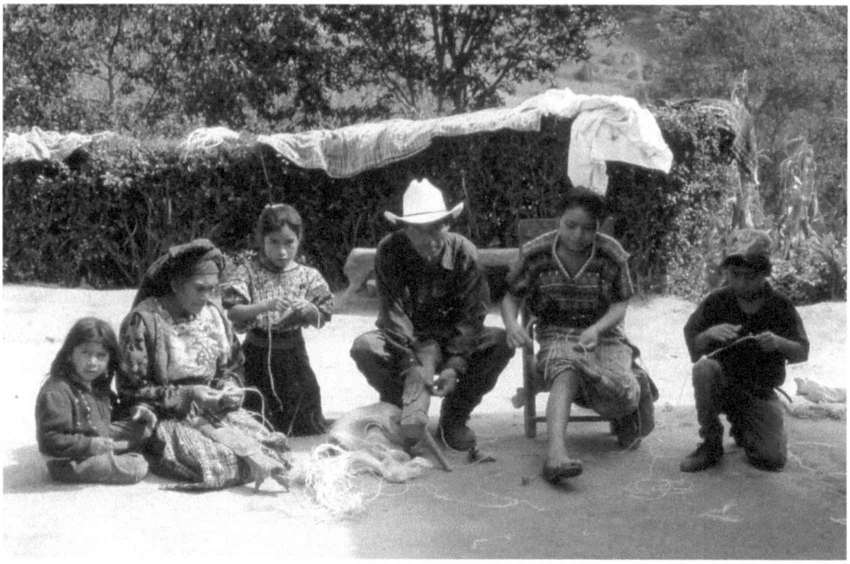

Figure 9.1 Ermiño (with palo) and family members leg spinning, Concepción Tutuapa, San Marcos.

knew I would pay in cash. Unfortunately, the hen died before she laid any eggs and they weren't sure if the price of the palo should be based on the hen alone or the unborn chicks as well. I let them figure it out and then paid their asking price.

While a palo is found in most households, some younger women prefer spinning with a piece of tire or the bottom of a rubber sandal on their lap, which is held in place with an attached string tied around the spinner's waist. This is how spinners use their leg in San Marcos la Laguna, Sololá, and Chichicastenango, El Quiché. Here, men are the primary spinners, and they place a piece of tire, leather, or shoe bottom on their upper thigh and tie a string around the upper leg, with a loop that is attached to the toe to keep it in place (fig. 9.2).

Men in San Juan Cotzal also spin on their leg, and they agree that this is the best method for morral making. Wet pine pitch placed on a shoe sole, inner tube, or leather tied around the thigh is common in Cotzal.

Hand Spinning

In San Juan Chamelco, San Cristóbal Verapaz, Santa Cruz Verapaz, Alta Verapaz, and San Andrés Huista, Huehuetenango, spinning is accomplished

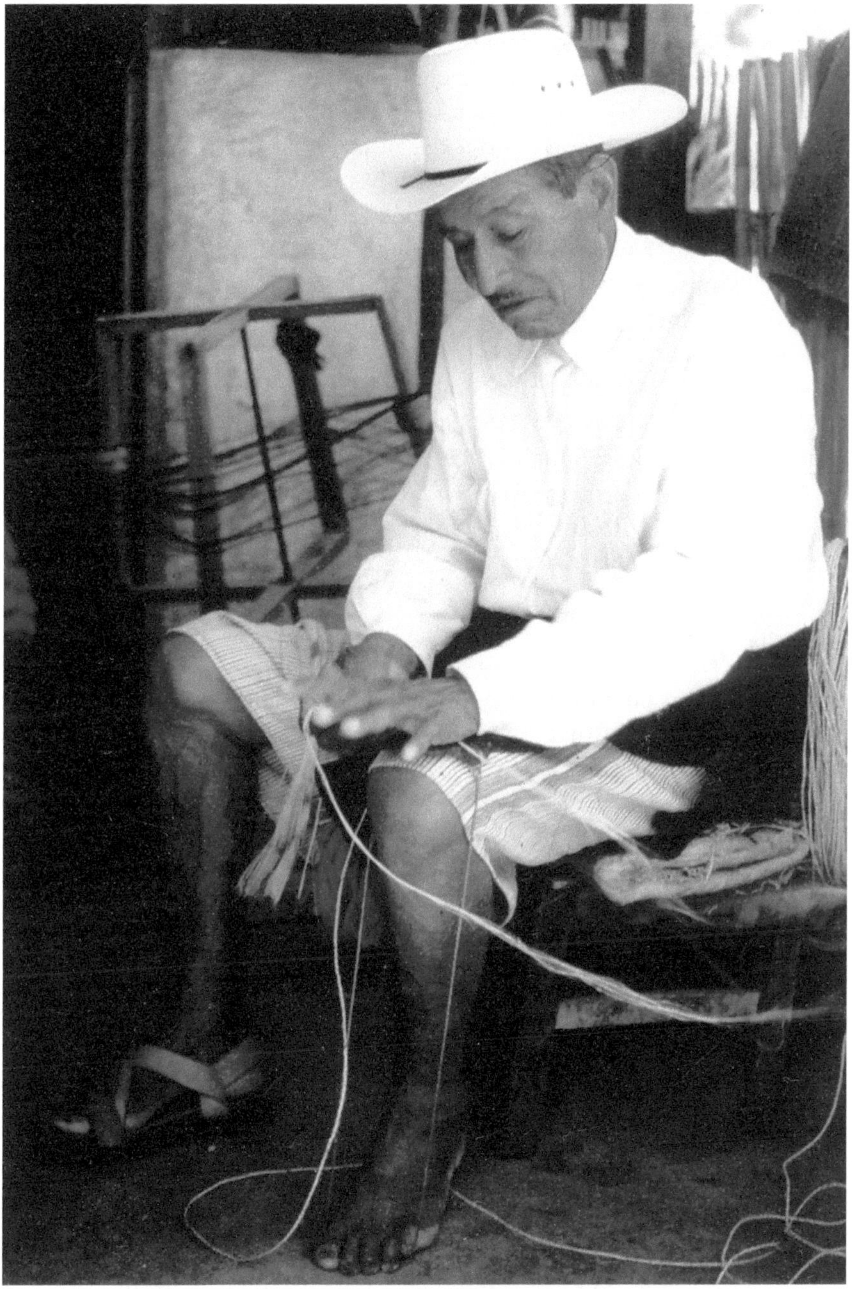

Figure 9.2 Lucas leg spinning, San Marcos la Laguna, Sololá.

between two hands using the same motion as leg spinning, both twisting and plying the strands together. One woman who demonstrated this process to me explained that her children always said that spinning was "magic," because they saw one quick motion of unspun fibers going in one side of her palms and coming out the other side spun. The process looks easy when performed by someone who has been doing it for many years, but, as a novice spinner, I found it almost impossible to get a decent spun strand, even after months of leg spinning. The best time for spinning is when it is misty out, as the moisture in the air helps the fibers adhere. Sometimes pumice or *comejen* (ashes of burnt termite tailings) is used to help the fibers adhere.

In San Andrés Huista, large amounts of pita are hand-spun in advance for linked bags. An advantage to leg and hand spinning is that these techniques do not require much space, can be done alone, and can be performed anywhere. A couple of men mentioned doing leg spinning during the rainy season, when they can't spin outside with other methods.

Taravilla

The taravilla, or tarabilla, is a common handheld apparatus consisting of a heavy piece of wood that spins around a handheld dowel. Maguey fibers are attached to a notch in the wood and, as the spinner quickly rotates the wood slab, another person feeds in new fibers, walking farther from the spinner until the desired length is reached. The taravilla was an important tool on sailing ships for rope-making emergencies in the 1500s (Hopkins 2001, 21). Taravillas could have traveled to Guatemala on these ships, but they were known throughout Central America prior to outside contact (Girard 1949, 263).

Taravillas made from guachipilín wood are a common sight in San Pablo la Laguna, Sololá, where women can spin about two hundred *varas* (a vara equals approximately one yard) in one day (photo essay fig. 5). To produce singles, one taravilla is used, and for plying, two are used (one in each hand).

Other locations where I observed taravillas are Olopa, Comapa, San Cristóbal Verapaz, and San Juan Sacatepéquez, where taravillas are made from *conacaste* (*Enterolobium cyclocarpum* [Mimosaceae family]) wood, and in Acatenango, made from guachipilín wood. Some are made from oak.

Spinning Wheels

Large spinning wheels, introduced by the Spaniards, are found all over Guatemala. I am uncertain of the origin of small spinning wheels. The large wheels usually have two hooks per wheel, whereas the small wheels are either one wheel and one hook or two wheels and two hooks. Spinning is much quicker with a wheel than by hand or on the leg, although the completed strand is usually not as tightly spun, which could be one reason morral makers prefer leg spinning.

The *cirenuela* or *torno de mano* (hand twister) is one or two small (five-inch-diameter) wheels situated in a handheld wooden frame, with one or two hooks where the strands are attached; they produce one- or two-ply strands (fig. 9.3). The hooks rotate by the downward motion of the hand on the center dowel. I saw them in San Pedro Pinula, Jalapa; Chichicastenango, El Quiché; and San Juan Sacatepéquez, Guatemala.

In Jocotán and Olopa, Chiquimula, and San Andrés Huista, small (six-inch-diameter) wheels are held between tall sticks. The wheel turns a hook, which is where the fibers attach. In Olopa and Jocotán, they are "planted" in the ground, whereas in San Andrés Huista the wheels are portable and taken out only when needed, which could be because it rains more there. The wheel is spun using the bare hand or by wrapping a string around the inner dowel and pulling it back and forth (photo essay fig. 6).

I saw large spinning wheels in San Juan Sacatepéquez, Guatemala; San Juan Chamelco, Santa Cruz, and San Cristóbal Verapaz, Alta Verapaz; and San Juan Cotzal, Quiché. Most were made from wood, but a few used bicycle wheels instead of a wooden wheel.

Carretas are large, round, wooden spinning wheels with a handle protruding from the middle. Fibers are attached to metal hooks, and as the handle is turned, two helpers "feed" fibers into the spun end, walking farther and farther away down the street (fig. 9.4). This action creates long strands of pita that are wound into one-pound bundles.

In Comapa, women and children spin maguey fibers on beautiful *caoba* or mahogany (*Swietenia macrophylla*) wheels and they use small wheels or *contratornos* (counter-spinners) to ply.

I did not see any large spinning wheels while in Jocotán, but they were in use when Wisdom visited (1940, 151). In 2009, I observed three bicycle spinning wheels in the Olopa AMO office, but they were not being used because they would not spin in reverse, a necessary spinning action.

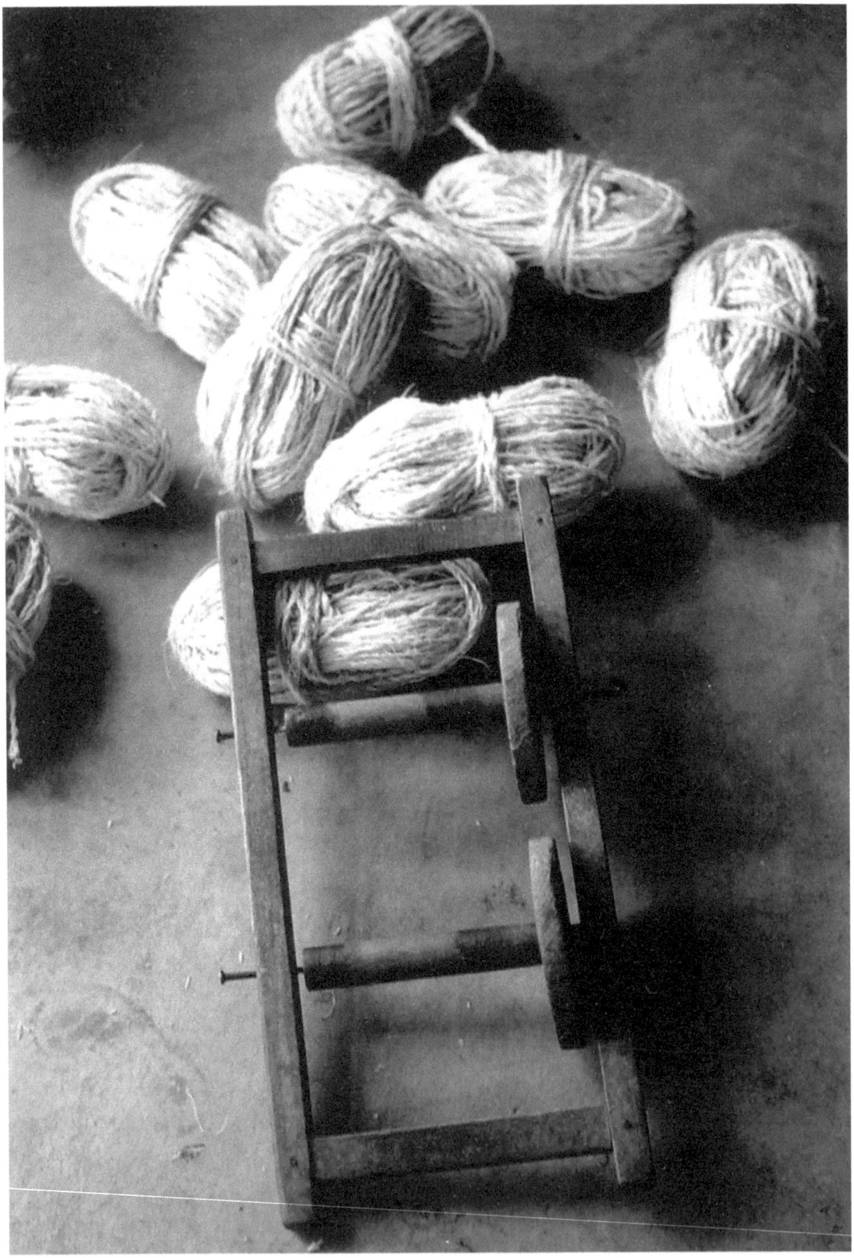

Figure 9.3 Cirenuela (with two wheels) with bundles of spun maguey, San Juan Sacatepéquez, Guatemala.

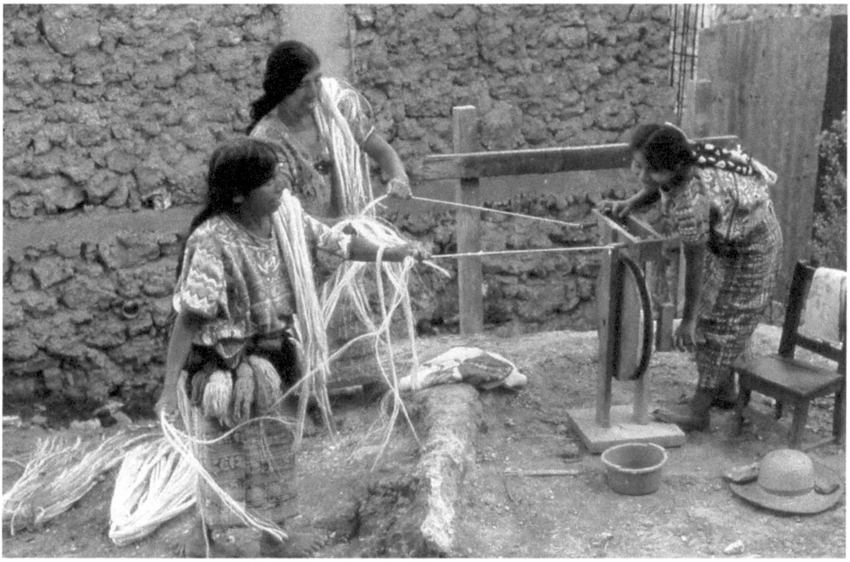

Figure 9.4 Isabella and her helpers with a spinning wheel, San Juan Cotzal, El Quiché.

In Concepción Tutuapa I heard of three "spinning machines," and after locating them, I found them to be identical to spinning wheels that I saw in other areas.

Taravillas, cirenuelas, tornos de manos, and spinning wheels are used in locations where a large amount of pre-spun fiber can fit through the working space, such as when making loom-woven bags, hammocks, and horse gear. Spinning with these tools is always done outside, as a long, straight, unobstructed space is crucial to the work.

In central Mexcio the Otomi spun fine strands of maguey with spindles for backstrap loom-woven *ayates* (carrying cloths) (Parsons and Parsons 1990, 180–181; Granberg 1970, 14). Spindles are not currently used with maguey in Guatemala, but they might have been used when fine maguey cloth was woven.

Hand-Cranked Rope Spinner

Rope can be many ply and thus an apparatus with more than two hooks is necessary. I saw a four-hooked large wheel in Concepción Tutuapa, where it was shown to me by a woman recently relocated from Sipacapa, which is known for rope making (fig. 9.5).

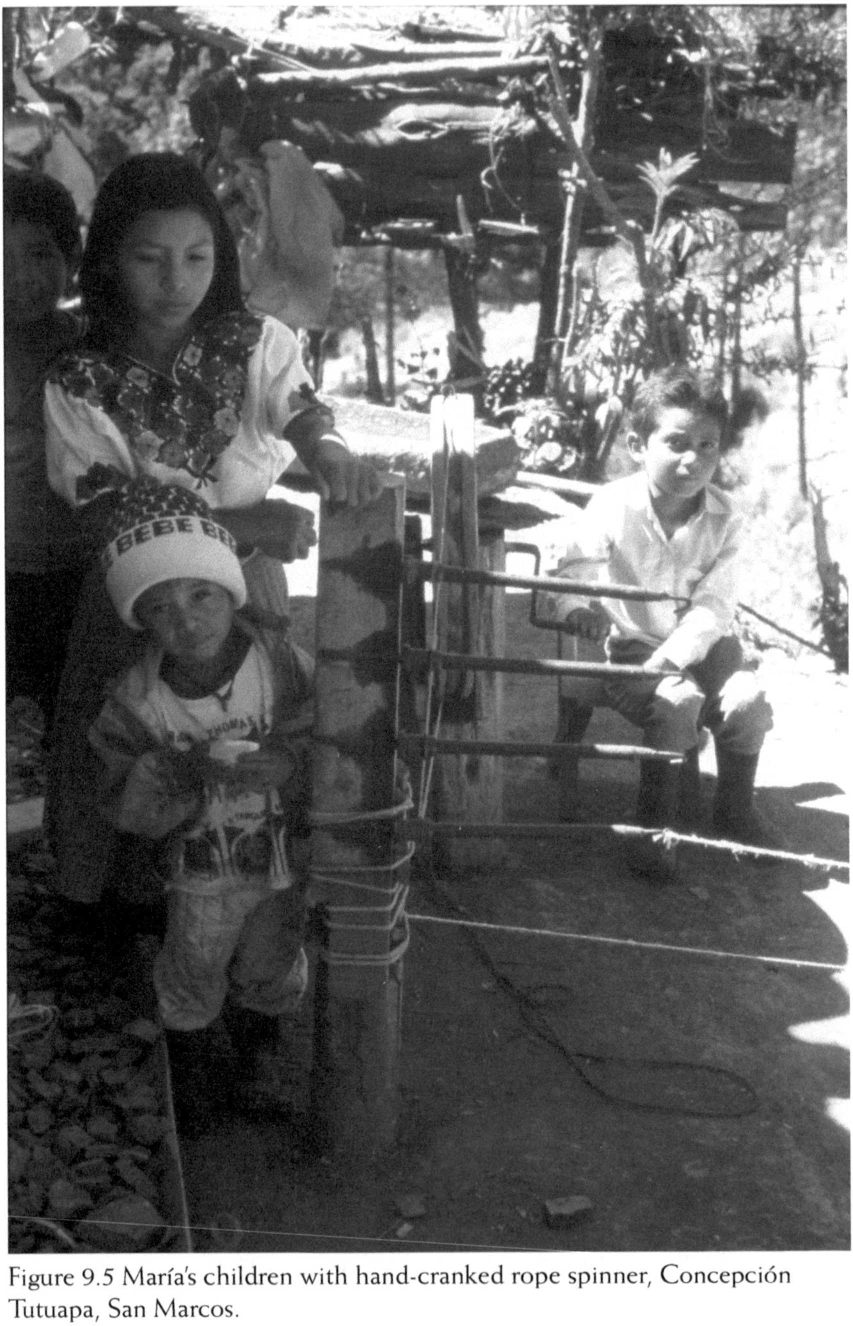

Figure 9.5 María's children with hand-cranked rope spinner, Concepción Tutuapa, San Marcos.

Electric Rope Spinner

The electric rope spinner is uncommon, most likely because of the cost to purchase and operate one. The day I observed its use, one-and-a-half-inch-thick maguey rope was being spun. Five women fed fiber into a strand that was spun by the electric spinner. They used two pounds of maguey for each length of 220 yards. After walking back and forth fifteen times, the strands were spun tightly and, when a counter-spin was let loose, a thick rope about 165 yards long was created. One interesting addition to the rope is that two strands of hand-spun maguey are always placed longitudinally in the center, which people like even though they don't serve a purpose. The rope produced for ranchers has diesel added to the fibers, said to make it waterproof.

After the fibers are spun into strands, the work of making items can begin. It is rare to find quantities of pre-spun maguey for sale anywhere in the country, with the exception of small bundles, which are used for attaching the warp threads to a backstrap loom. They are sold where weaving equipment is found in the Chichicastenango market.

10

From Strand to Product

Maguey items are made using non-loom, or hand-constructed, and loom-woven techniques. Some methods are very portable and do not take much working space or tools, and can be made outside or inside. Others require large tools and space and are usually worked outdoors. As with all handiwork, it is always obvious if more than one person has worked on the same project. Some techniques most likely developed in the region, whereas others were introduced.

Non-Loom or Hand-Constructed

Looping

Looping, often referred to as knotless netting (along with a variety of other names, including nälbinding, which is Scandinavian for "needle binding"), is an ancient and common single-element bag-making method. Looped bags are also made in Mexico, El Salvador, Panama, Columbia, Ecuador, Peru, Paraguay, Australia, and Papua New Guinea. There has been great debate as to whether looping developed independently in each region or migrating peoples brought the knowledge with them (Davidson 1935, 117–134).

Looping is achieved when a single strand crosses over itself to form a loop, and the end of the working strand goes through the loop, which is why a large amount of pre-spun material cannot be used. MacKenzie suggested that looping is one of the earliest textile processes known because working with pre-spun fibers is not possible and thus tools were not used (1991, 210). Nordland agrees that looping most likely developed before the invention of spindles and whorls (1961, 134).

The structure builds by spiraling and building loops on each successive row. Looping a square bag begins with a row the length of the desired width. For a round bag, a single loop is made, and loops are added to the original loop in a circle. From these loops more are added, increasing its width and eventually the desired height. To decrease the size, loops are skipped. The size of the loops depends on the desired use of the final product—bigger loops create a more elastic bag. Tools are not necessary, but I find a blunt-pointed needle with a large eye helpful when working on bags with small loops. In Jocotán, workers use shuttles to make cargo nets, probably because they are comfortable using them to make hammocks. The different varieties of looping found in Guatemala are as follows.

Simple loop or buttonhole. Simple loop (Emery 1966, 31) or buttonhole (Seiler-Baldinger 1994, 11) is the simplest and possibly the oldest looping stitch (Nordland 1961, 141). In Guatemala, most simple-looped morrales are begun with a single loop and a subsequent row is made to the desired width with figure-eight looping (see "Figure-Eight or Hourglass" below). The trickiest part of the bag is turning the corner to create the second row of eights that forms the bottom, because the strand goes through two loops instead of one, and then the entire work is turned in the opposite direction to be worked back down the existing row. After these two rows are complete, single looping continues row after row, until the bag is square. The final row is slightly different, with the strand going through the last two loops, instead of only one, creating a firmer edge.

Maintaining equal stitch tension is important; otherwise, the bag will "pull in" and not look good, for the body of the bag is made without the use of a mold or form. Bags are often made inside out so that they remain clean on the outside. Because looping is alternated with leg spinning there are no knots, making these bags extremely strong and durable, for knots are known to weaken the fiber (Edholm and Wilder 1995, 1). After the body of the bag is complete, the same strand continues to form the long strap. I found that novice workers often do not know how to construct the strap (see "Interlacing" below) and will spin enough material for an expert weaver to finish.

After the strap is complete, the final step is attaching a loop to the opposite side of the bag. This is the only added element of the entire bag and is important because the long strap ties off to it, forms the handle, and enables the wearer to adjust the strap length. Completed bags are sprayed with water, blocked, and dried in the sun. Simple-looped bags use sixty to eighty leaves of maguey and take about a month to make. They are found nationwide and are the most common style in Concepción Tutuapa.

Single-interconnected or simple-interconnected loop. Bags made using the single-interconnected (Emery 1966, 33) or simple-interconnected (Seiler-Baldinger 1994, 13) technique were seen in Concepción Tutuapa and Jocotán. They are constructed using a series of small loops, which are not pulled tight, but interconnected in an open manner. In Guatemala I saw only plastic bags made with this technique, but when I went to El Salvador I found maguey bags of this method. In 2009 I happened to see a maguey single-interconnected bag in the Chiquimula market, and I would guess, it was imported from El Salvador.

Figure-eight or hourglass. Figure-eight (Emery 1966, 33) or hourglass (Seiler-Baldinger 1994, 14) is another looping technique. In Concepción Tutuapa, small, square bags are also made using figure-eight looping, which produces an open, stretchy bag called a *cachelpa* (matate). In spite of the basic construction, it can be very difficult to make the "eights" all the same size. After completing a small bag of this style, I proudly showed it to a woman who immediately grabbed it and took it home to rework the entire piece so that every row was uniform.

In San Juan Cotzal, matates are also made using figure-eight loops. In contrast to other locales, the matate is worked while attached to a strand tied around a man's thigh, which is removed when the bag is complete.

Large figure-eight looped round bags or *redes* come in many sizes, but the ones that fit around market-sized bamboo baskets are the most popular. In San Juan Cotzal I was given a two-hour demonstration of how to make one.

The bag was worked around the legs with spiraling figure-eights so that, when complete, the maker's legs were completely enclosed. To begin, the first row of loops was placed on a single strand of spun maguey that was wrapped around the weaver's thighs. When complete, this strand was drawn tight, which created the bottom of the *red*.

Osborne described Nebaj *red* making in much the same way in the 1960s: "The weaver begins his bag down at his ankles and, working around his body, continues upward until the top edge of the bag is reached near his waist. He then steps out of the finished bag and inserts strings in the loops at the top and bottom to close the ends" (1965, 177).

Cotzal redes come in two sizes, determined by the total number of loops and rows. A red for containing corn has fifty loops that are three fingers wide (about two inches); a *red* for pottery has forty loops that are six fingers wide (about five inches). Both are eleven rows long. To make a standard-sized *red*, ninety yards (one and a half pounds) of fiber are needed.

In other locations redes are not made around a worker's legs, and in Jocotán, redes do not spiral as in other locales but are built one row at a time. They begin with twenty loops, double in the second row, and so on. The preferred redes are smaller than the ones in Comapa. Here, ten leaves' worth of fiber (one manojo) is enough to make a *red*. Two redes can be made in one day. Santa Cruz Verapaz redes are made the same way as in Jocotán, with the first row going around a small loop, and doubling the number of loops in the second row. The *red* is begun around a spun strand, much the same way as I observed in San Juan Cotzal. After redes are made they are washed and dried in the sun. I never saw a *red* made with any other technique (photo essay fig. 7).

Cross-knit or encircled looping. Cross-knit looping (Emery 1966, 32) or encircled looping (Seiler-Baldinger 1994, 17, 25), known as *cadena* (chain) in Guatemala, looks like knitting but the end of the working strand goes through each loop, every stitch, whereas in knitting it does not. If a knit bag is cut, it will unravel, unlike a cross-knit looped bag, which won't. The actual stitch is similar to simple looping, except the loop placement is around the crossing of the loop of the previous row and not through the larger space between the loops.

This technique is usually tightly done and produces distinct vertical lines, thus the name "cadena." It is very tedious and uses much maguey, which is probably why few people make cross-knit bags today. I was lucky to meet a man in San Juan Cotzal who does, as the only other cross-knit looped bag I found was in Jacaltenango, where a woman selling produce in the market had one. She told me the name of the maker and that he lived in a Colotenango aldea, once known for morrales (Wagley 1969, 65), but I never found him, even after an extensive search.

Knitting

Knitting (Emery 1966, 40) or vertical interlooping (Seiler-Baldinger 1994, 24) is a single-element technique where loops are interlocked in parallel rows using sticks or needles. Knitting possibly originated in the Middle East or northern Africa and was likely introduced to Guatemala.

The bags of Chichicastengo are made with a circular needle and are knit in the Portuguese method by "throwing the yarn with the left hand, keeping the left-hand needle inert and activating the right-hand needle" (Rutt 1987, 203). The knitters of San Marcos and San Pablo la Laguna use three tapered sticks in a method I define as "finger-knitting," where the worker's fingers, not the needles, move the loops from one stick to the next (photo essay fig. 8). The tapered sticks are not pointed and are of approximately the same size because they are size gauges or calibers for the loops. O'Neale describes this technique, although she doesn't say where she saw it (1945, 183).

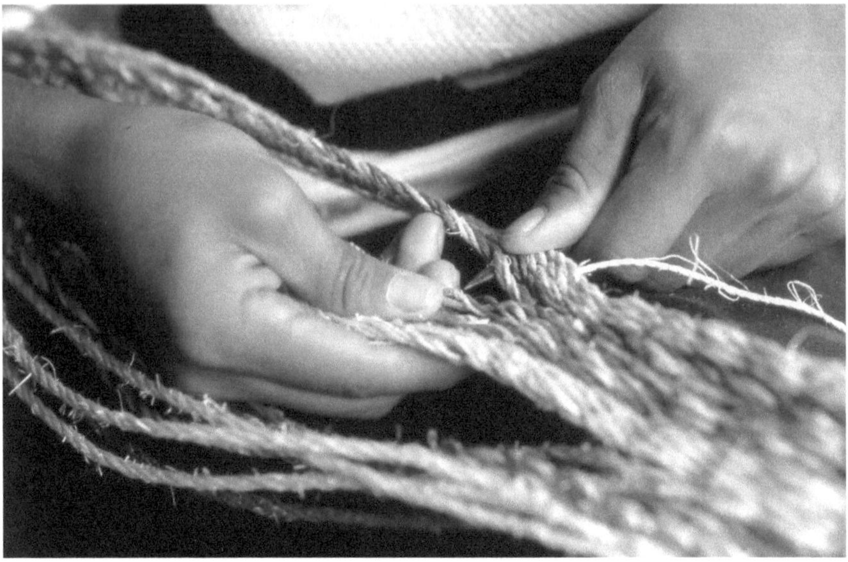

Figure 10.1 Close-up of ply-split darning, San Pedro Pinula, Jalapa.

As with looped bags, the bottom row determines width, and the rows spiral until the desired height is reached. If a final stitch to lock the last row into place is not done, knitting will unravel.

Ply-Split Darning

Ply-split darning (Collingwood 1998, 13) is a technique that is documented for making camel gear in India and horse gear in South America (Collingwood 1998, 200–201). In Guatemala it is also used to make horse gear (tenedores and gamarrónes) and, because of this, I would guess that it arrived with the Spaniards because they introduced horses to Guatemala. A long needle is used to split the ply (twist) of the spun strand, pulling a separate strand between (fig. 10.1). I observed the process in an aldea of San Pedro Pinula, where two young brothers demonstrated the making of a tenedor.

First they spun damp fibers, some colored with aniline dyes, with a cirenuela. Using a warping board, strands were wrapped around pegs and then removed. Then the entire piece was placed on a frame, with the strand ends looping around wooden posts at either end of a piece of wood. A separate strand was placed between the ply of the warp strands with a five-inch needle, and ply-split darning was worked across the entire width for two inches. Then the strands were split into two parts, and the same process continued down both sides. When complete, the tenedor was removed from the frame and wet *sepa de guineo* (the "fabric" or sheath of a *guineo* or banana plant) was wrapped around one end, covered with damp leather, and sewn shut. The entire production took about three hours.

Ply-split darning is also used to make gamarrónes, which I saw in Comapa. First, many different thicknesses and lengths of rope were spun with wheels and taravillas and then interlocked using the technique of ply-split darning. When complete, the gamarrón was painted with liquid aniline dyes. It took about two hours to make one.

Ply-Split Braiding

Ply-split braiding (Collingwood 1998, 13) is similar to ply-split darning except that the ply is split diagonally instead of perpendicularly. Most likely the ply-split braiding method also arrived with the Spaniards because, as with ply-split darning, documentation of horse gear made using this technique in South America confirms this (Collingwood 1998, 18–20, 200–201). The only example I saw of ply-split braiding in Guatemala was on the clews, or ends (eyes), of a Jocotán hammock. It was created in a local prison where inmates are often taught techniques for making items like hammocks, and I do not know where the instructor learned ply-split braiding. In 2009 I saw a beautiful maguey hammock in San Salvador's Museo Nacional de Anthropología David J. Guzmán that looked to me like the very ornate clews were also made from ply-split braids.

Linking

Linking (Emery 1966, 30; Seiler-Baldinger 1994, 7–10) is the method for creating matates in San Andrés Huista and hammocks nationwide (Drooker 1981, 51, 52). Hammock linking techniques likely originated in the upper Amazon River basin of South America and spread via trading ships around the world (Drooker 1981, 3). The Arawak people made hammocks on frame looms, and similar loom designs were possibly introduced into Guatemala; however, the "looms" used for hammock making are not frame looms but two poles stuck in the ground, placed at the desired width of the hammock, usually outside (fig. 10.2).

The matate "loom" imitates that used for making miniature hammocks and is a two-and-a-half-foot-long board with metal rods sticking up at each end. The metal rods fit into holes, of which there are many spaced along the board, for different sizes of bags. When in use, the frame is tied down to the worker's legs with a piece of rope. A length of bamboo is bound at the top of both metal rods to keep them from pulling in, for as the bag grows upward, it has a tendency to tighten (photo essay fig. 9).

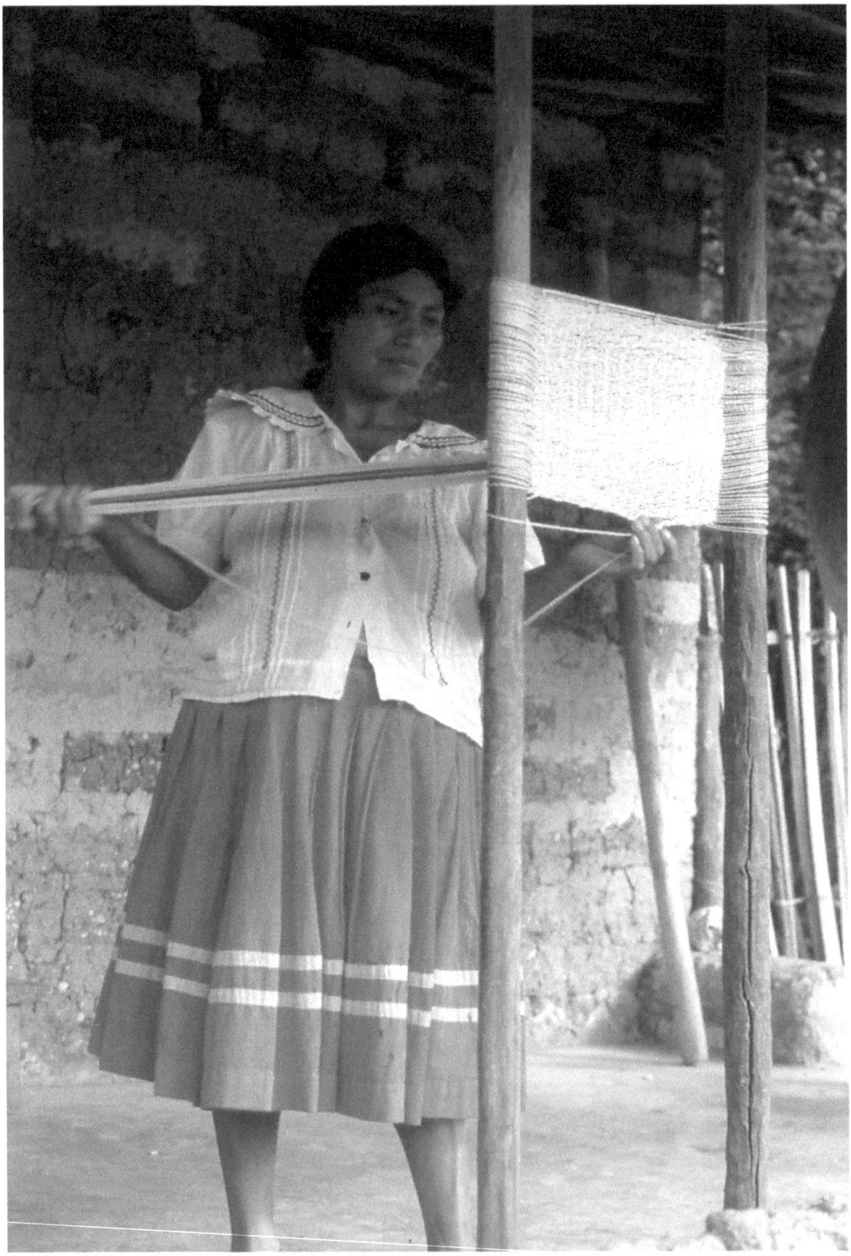

Figure 10.2 María making a small hammock, Jocotán, Chiquimula.

Matates are made using the single-element technique of simple linking, formed by consecutive overcasting or whipping stitches. To begin a matate, a single strand is wrapped around the metal rods and then, for example, twenty-two loops are added around the single strand. The strand is worked back and forth, alternating between loops and pulled tight. One side is worked, removed from the frame, and then as the other side is made, it is connected with the first side as it grows upward. Because of how linking is done, the bag is constructed sideways. The completed bag is removed from the frame;, the sides are sewn shut, and hand-spun thick rope is added for the strap and ties. A small matate uses about a pound of maguey and takes about fourteen hours to make; larger ones use two pounds of maguey and take up to three days to complete. To hold the final shape, completed bags are sprayed with water and dried in the sun for two hours.

Hammocks, on the other hand, are made with complex linking, where two single elements interact. The working threads are wrapped on shuttles, which are used to put the strands through the links.

In Comapa, four rows of *botons* (buttons or knots) are made first, which help keep the edges firm. Next, row after row of linking is worked and then finished with another four rows of botons. The hammock is made from bottom to top with the Guatemalan technique (see next paragraph), which forms the bed (the part that one sits or lies on) of the hammock. When complete it is removed from the frame and thick ropes are threaded through the warp end loops to form the clew.

In Jocotán, hammocks are made using two different linking variations. The "one pass," or *malla salvadoreña* (Salvadorian net), is defined as 1 3/3 (Seiler-Baldinger 1994, 9) because the working strand skips one row between link points and then travels over and then under three threads. The more common Guatemalan version is "two pass" or 2 22/22 (Seiler-Baldinger 1994, 9).The "one pass" technique uses more maguey than the "two pass" does. Hammocks are made in many sizes, based on the number of turns and stitches. First, five rows of botons are made, then, for example, an average-sized hammock will have seventy-five turns of twenty-six *mallas* (links). Five more rows of botons are made at the end. For maguey hammocks, a candle is rubbed on the wooden shuttle to keep it smooth and help it pass through the loops without catching on the "hairs." In Jocotán, hammocks are worked from top to bottom.

Crochet

Crochet (Emery 1966, 43), or vertical and lateral interlooping (Seiler-Baldinger 1994, 23), is a single-element method where the loops lock laterally as well as vertically. The word *crochet* is French for "small hook," which is the tool used (Rutt 1987, 10). Wool, cotton, and maguey bags are made with this technique, although until 2009 I did not see any maguey bags made entirely from crochet, only a single row to begin a morral in Tutuapa. Now, a few members of AMO in Olopa are using their crocheting skills to make maguey bags and decorative flowers.

Interlacing

Interlacing is the most common strap-making technique for looped and knit bags. When the desired bag height is reached, the same strand continues to create the strap. O'Neale describes the strap configuration as single-element "weaving" (1945, 183) and refers to Graumont and Hensel's pyramid knot, classified as an ornamental knot (1970, 320–321). The Graumont and Hensel diagram shows a knot shaped like a pyramid, hence the name. After I learned how the strap was constructed, I realized that the pyramid knot is really just the beginning. After a "knot" is created, the strand continues, weaving in and out, in a form of single-element interlacing, or "weave," where the same strand is both warp and weft, until the desired length is reached (fig. 10.3).

Interlacing is always worked away from the artist. One of the most difficult aspects of this strap process is retaining sufficient working space (in the center), as there is a tendency to pull tight. Also, once the loops begin to "climb" up the strap, pre-spun strands can be used, as the working end no longer passes through a small space. To end the strap, the single working strand is pulled through the final loops and knotted. In some cases the last loops are evened up and wrapped with a thin maguey strand, to form a single loop. Some interlaced straps are loosely made, and others are very tight and stiff. Usually interlacing is made using a continuation of the same strand from the bag, but in San Juan Cotzal I was shown an interlaced strap made as a separate piece and attached to the bag after it was complete. It is easy to mistake this technique as sprang (see next section).

Sprang

Sprang is a technique in which parallel fixed-warp threads are manipulated by interlinking, interlacing, or intertwining from one side of a warp (Collingwood 1974, 31). Because it is fixed at both ends, the same structure

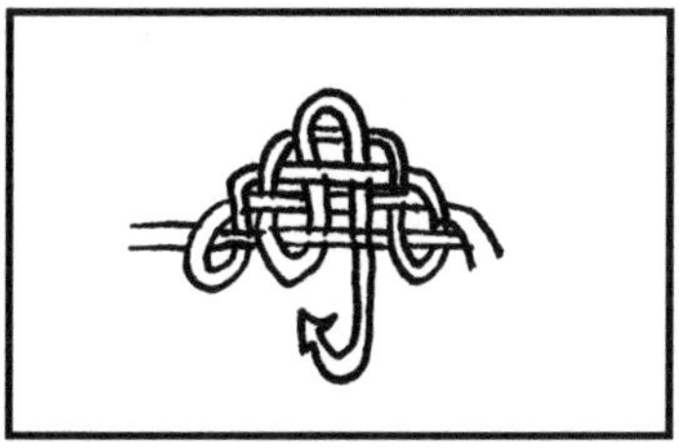

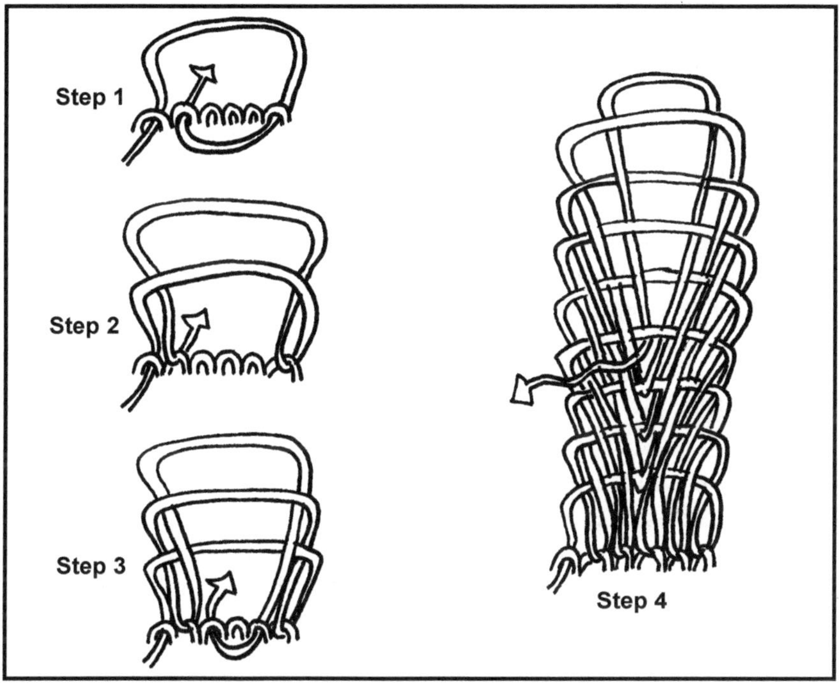

Figure 10.3 Diagrams of a pyramid knot and interlacing. The most common strap for a looped or knit bag is made with the technique of interlacing. The first step is to create a pyramid knot and then continue with a single-element weave, where the working strand is both the warp and the weft.

occurs in reverse on the opposite side. When the work meets, something must be placed or tied at this point to prevent the entire piece from unraveling (Emery 1966, 66; Seiler-Baldinger 1994, 51). The warps between saddlebags are manipulated in an adaptation of interlaced sprang (Seiler-Baldinger 1994, 53) (see appendix 3).

After the bags are woven at each end, with the warps under loose tension, the unwoven strands are braided. Simple saddlebags have braids only, and fancy bags are plaited as one unit for about two inches and then the warps are divided into three-strand braid groupings. A single strand secured perpendicular to where they join keeps the warps from loosening or unraveling (O'Neale 1945, 182; Collingwood, 1974, 278).

Braiding or Plaiting

Braiding with multiple strands (Emery 1966, 63), or oblique interlacing (Seiler-Baldinger 1994, 39–40), is part of making a saddlebag (see above) and is sometimes used for making straps on looped bags. This type of strap can be woven separately and attached to the bag after it is complete, or begun on a completed bag. This is a common strap technique in San Juan Cotzal, Panama, and El Salvador. Another strap style is intertwined, beginning about two inches below the top of the bag and then "finger-woven." O'Neale also describes bag straps as being "finger-woven" in San Juan la Laguna, where a man manipulated the weaving using only his hands (1945, 183). I saw bags with woven straps only in Jocotán, and only in plastic, and they might be coming in from El Salvador. Unwoven warp ends are also braided on loom-woven bags.

Macramé

Macramé (Emery 1966, 65), or interknotting (Seiler-Baldinger 1994, 47), is based on plaiting or knotting (Graumont and Hensel 1970, 327). The word comes from the Arabic *mucharram,* meaning lattice. Macramé is closely related to oblique intertwining, except that looping and interknotting are combined (Seiler-Baldinger 1994, 47). This technique is uncommon in Guatemala, and the only two macramé bags I saw are in the Museo Ixchel's collection.

Plait Coiling

Plait coiling (Sentence 2001, 64) is a common hat-making technique. Handmade hats have plaited palm strips sewn together with maguey fiber or *pita de cosedora* (sewing thread). In 2002 one man in San Pedro Pinula still made them by hand, while everyone else used a sewing machine (and cotton thread). Because of the historical use of maguey fibers, I include hat making.

The palm used for hats in San Pedro Pinula is bought from vendors of low-lying San Jacinto, Chiquimula, and Asunción Mita, Jutiapa, because Pinula is too high and cold for *palma real* (royal palm) to grow. The fronds are cut three days before the full moon (Osborne 1965, 194) and then split into strips, with the middle part sold to women who make brooms from it.

Women in the aldeas of Aguacate, la Ceiba, and Aguitos dye and plait the palm strips. Aniline dyes are common, but natural dyes are favored because I was told that they do not fade in the sun. The color brown is achieved by soaking the palm in mud for one to two weeks, black from boiling the brown-dyed palm with the bark of the *jocote* (yellow oak) tree, and yellow from boiling the palm with *palo rosa* (rosewood) bark. The mud can be used five times before all of the color is exhausted. One woman said that she only makes brown dye because firewood is too expensive to boil the bark needed for the black dye.

After the palm is dyed, it is plaited into strips six *abrazos* (one abrazo is about a yard) long. These are then sewn in the form of a hat, starting in the middle of the crown and spiraling around to the brim. As the strips spiral they are sewn to the previous row.

Plait coiling was taught to maguey workers in Jocotán for making baskets and placemats. Similar to hat making, the form is achieved by spiraling long three-strand maguey braids around and sewing each new coil to the previous coil, with a "stitch" of simple looping (Emery 1966, 45). Each basket takes three days to make, uses a pound of maguey, and sells for thirty quetzales. Baskets made from braided lengths are stronger than those made from loose maguey strands (Sentence 2001, 64).

Wrapping

Wrapping is a technique where loosely spun maguey is placed around an interior object to from a ball. In this sense it is not an individual stitch, but an entire process. This is how the *bomba* (firework, skyrocket, or other pyrotechnic) is made. I observed bomba making in several small *coheterías* (fireworks shops). To begin, two workers take one pound of maguey and loosely spin it using tornos de manos. These strands are put through a pot of *la brea* (warm pine pitch) and wrapped around a paper pouch filled with gunpowder (for exploding). A waxed metal spike is poked through the maguey mass, where a different type of gunpowder is placed (for shooting in the air), and a string wick is added. Sometimes chile pepper is included. There are five sizes of bombas. Maguey from Alta Verapaz is preferred over other regions, including El Salvador, because of its superior compression characteristics. A man can make twelve dozen in one day. Many municipalities have bomba workshops, for explosives are incorporated into a variety of events and ceremonies.

Loom Woven

Guatemala is well known for its woven fabric, and looms are found nationwide. Backstrap or stick looms have deep roots in the region, and Spaniards introduced the treadle loom and possibly frame-type looms. During colonial times there was an extensive trade network of woven cotton for the elite. Commoners were prohibited from wearing cotton and were required to don maguey clothing (Fowler 1989, 110); their lower-class male ancestors were also prohibited from wearing cotton and wore *mantas* (all-purpose, rectangular-shaped fabric) and loincloths made of agave, yucca, and palm fibers (Anawalt 1981, 27, 100). Spinners perfected their skills to produce very fine maguey threads that could be woven into pliable cloth and fashioned into mantas and huipiles (Osborne 1965, 11). Maguey clothing is no longer made, but maguey bags, horse gear, and mecapales are still woven on looms.

Loom-woven fabric has a warp and a weft and is a two-element technique. The warp is the structural component and is attached to the loom sticks, while the weft passes in and out of the warp at a perpendicular angle. String heddles, consisting of a continuous length of pita, are made on the completed warp of all looms except treadle looms, which have a rigid heddle, usually made of metal. The heddle pulls alternative warp threads up and down to create a shed, or space, where the weft threads lie. Woven maguey products are warp-faced, simple, or plain-woven lengths of cloth (Emery 1966, 76; Seiler-Baldinger 1994, 87). The weft is always spun softer than the warp.

With the exception of the treadle loom, maguey warps are put on the loom the size of the finished piece and are usually not cut. Often weavers have a cup of water nearby and will spray water from their mouths onto the maguey threads, to soften them. Beaters are kept smooth by rubbing paraffin wax on them, which prevents loose "hairs" from catching on the wood when beating. Sometimes wax is rubbed on the warp threads while weaving, also to make it more pliable.

In some areas maguey fibers are used only as weft material for finely backstrap-woven cotton belts. I saw examples of these in San Pedro la Laguna, Santiago Atitlán, and San Marcos la Laguna and later learned that, on the day I met Lucas en route from San Pedro to San Marcos (see chapter 3), he was returning from selling maguey to a belt weaver. He told me that one leaf's worth of fiber is enough for three belts. This is not a new technique. Gustavus A. Eisen collected belts of this type from many highland communities in 1902 (Schevill 1993, 220); in the 1960s O'Neale saw them in Totonicapán, Huehuetenango, and San Pedro Sacatepéquez and noted that maguey weft made them more durable and created "a strong belt with a good body" (1945, 158). O'Neale also discovered head ribbons woven with a maguey weft in Totonicapán (1945, 20).

Backstrap Loom

The backstrap loom is composed of sticks and rope, with a strap connected to the bottom end of the loom, which goes around the weaver's back. By moving his or her body forward or back the weaver controls the warp tension. Moving forward lessens the tension and opens the shed, whereas leaning back tightens the warp, allowing the weaver to beat the cloth. The other end of the loom is tied to a tree or house post and dictates the angle of the actual weaving. In some areas warp strands are measured on a warping board, before being put on the loom (fig. 10.4).

Concepción Tutuapa is one place where backstrap looms are used to weave mecapales and cintas (fig. 10.5). Mecapales are about twenty-two inches long and three inches wide, and cintas measure about two feet long and six inches wide. In takes approximately an hour to weave a cinta. Five cintas can be made in a day, and they sell for forty-five quetzales a dozen.

In San Juan Cotzal, mecapales are loom-woven by women, but all other maguey loom-woven goods are imported from other parts of the country. This could be because of a demand in Cotzal for backstrap-weaving straps, which are the same size as mecapales.

Fixed-Vertical Loom

Fixed-vertical looms are found only in San Juan Sacatepéquez; they are identical to a backstrap loom except that there is no backstrap. Instead, the bottom bar is pegged into the ground, and the other end is tied to the top of a tall pole that is planted in the ground (fig. 10.6). This modification, possibly invented by backstrap weavers, makes the weaving vertical and immobile and is probably easier for manipulating the stiff maguey fibers. Tension changes are achieved by adjusting the rope where it attaches to the top of the pole. Weavers using fixed-vertical looms often sit on a low stool to weave.

The woven bags have one selvaged or finished edge, which is accomplished by placing a thin metal rod or very small stick through the bottom end of the warp, allowing for the weaver to weave right to the end. The rod or stick is attached to the thicker loom stick with a strand of maguey that wraps around both. This is also a common technique used with cotton threads and is what the small maguey bundles found in the Chichi market

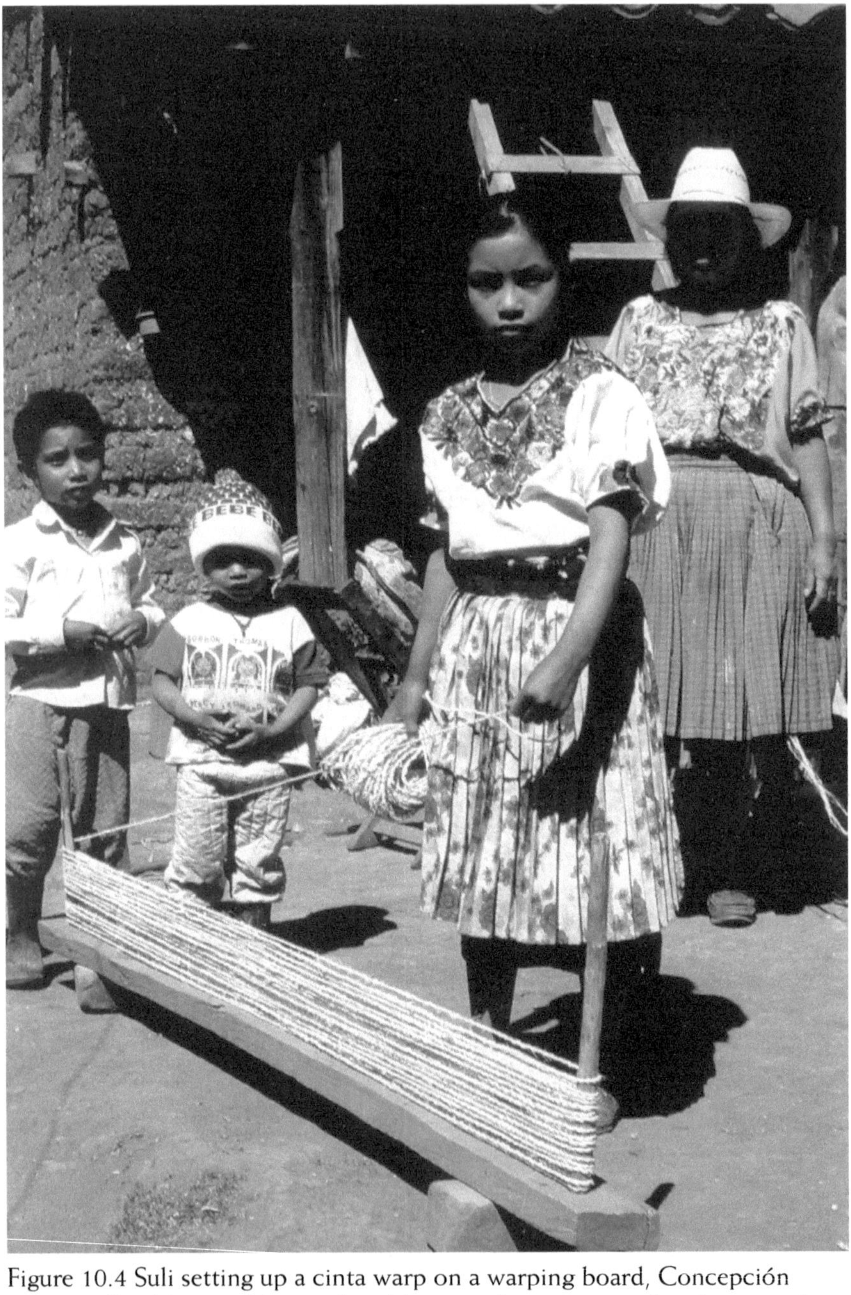

Figure 10.4 Suli setting up a cinta warp on a warping board, Concepción Tutuapa, San Marcos. Her mother, María, and two younger brothers stand behind her.

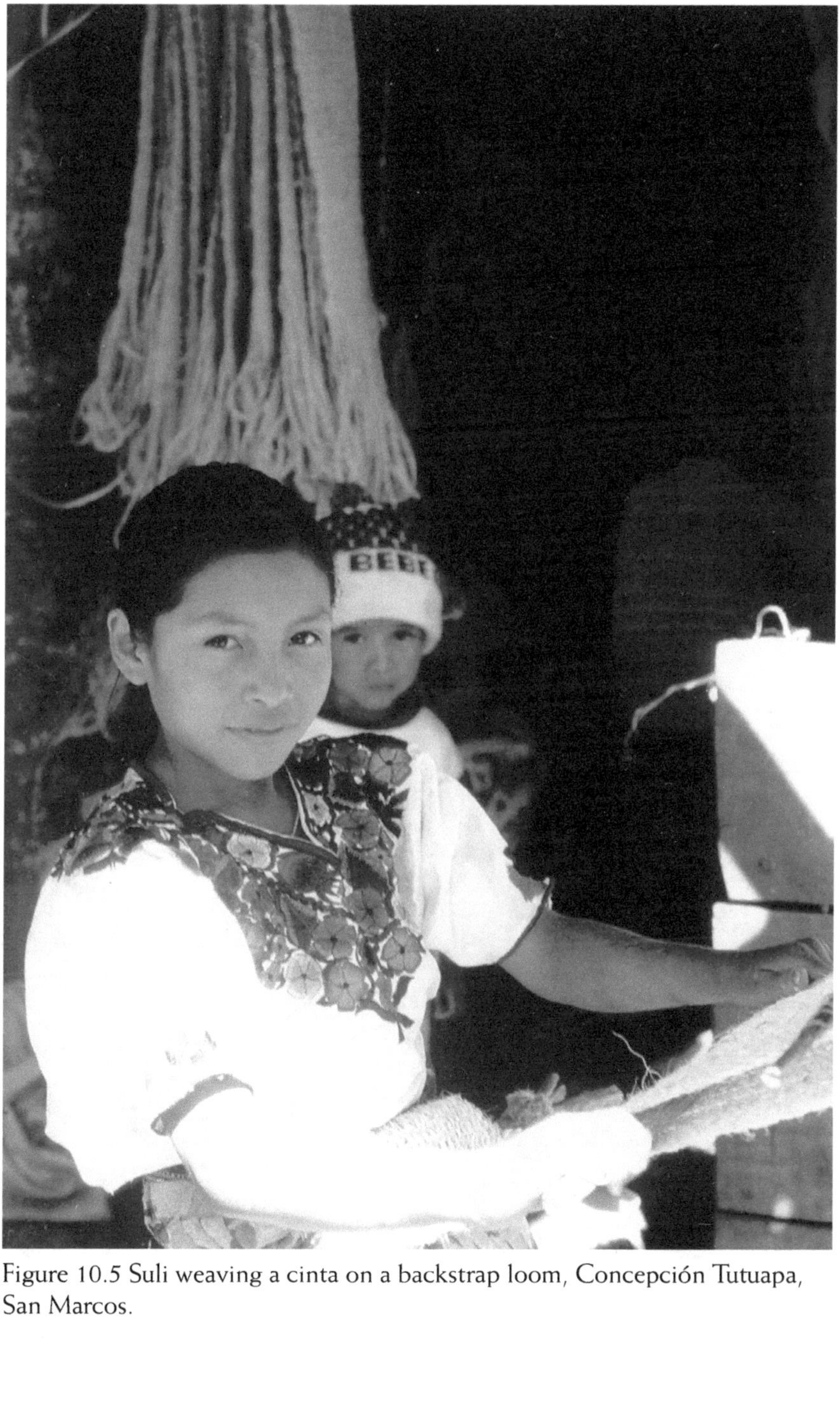

Figure 10.5 Suli weaving a cinta on a backstrap loom, Concepción Tutuapa, San Marcos.

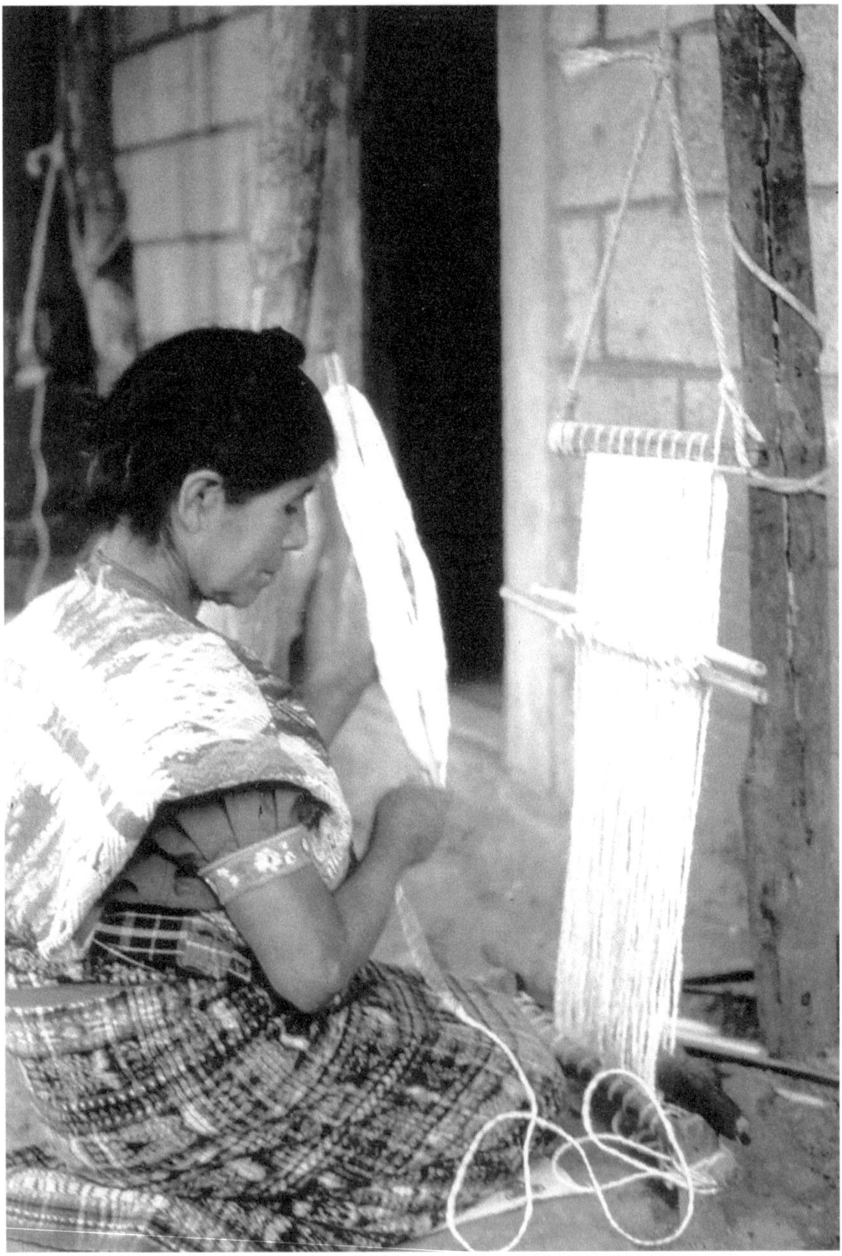

Figure 10.6 Basilia weaving a sembrador on a fixed-vertical loom, San Juan Sacatepéquez, Guatemala.

are for. The top three inches are left unwoven, and after removal from the loom, they are braided, which creates a decorative edge.

About seven leaves' worth of fiber and approximately one and a half hours are needed to weave a sembrador. Plain-woven bolsas are also made on fixed-vertical looms, and some have triangular top flaps, which are woven with decreasing warps until only a few are left. The narrow shoulder strap is also loom-woven, and after removal from the loom, it is sewn between the two completed side panels so that the strap forms the sides and gives width to the bag. Each bag takes a day to make and sells for thirty quetzales.

Saddlebags are woven as well and take about three days to make. Maguey mecapales are also made on fixed-vertical looms, and one entrepreneur had three set up at the same time. The looms shared the same heddle and beater, but were individually woven (fig. 10.7).

Vertical-Frame Loom

Vertical-frame looms were seen only in Comapa, although in 2009 I saw one outside the home of a development promoter in a Jocotán aldea.

The loom is made from two long poles, which are stored in the rafters until needed. When used, the poles lean against the outside wall of a house at about a 45-degree angle, with a distance between them of approximately two feet. Notches spaced for both saddlebags and bolsas are cut in the poles for the crosspieces. The weaver usually sits on a stool when weaving (fig. 10.8).

I witnessed several men weaving bags on the vertical-frame loom. Fancy bags use a pound of maguey and take about three days to make, which includes spinning, weaving, and finishing. The completed bags are decorated with braids on the front (fig. 10.9) and sometimes tassels or a fringe called *pico de gorrión* (sparrow's beak). In contrast, a bolsa corriente (plain or quickly made bag) can be completed in about forty minutes.

Horizontal Loom

Horizontal looms, found only in Olopa, are permanently planted in the ground, as either two forked sticks or four sticks. The weaver stands while working. Two crossbeams are tied across the ends of the four loom posts (photo essay fig. 10).

Unique bag styles made in Olopa are *glincas* (round bags) and *caristas* (rectangular bags). Glincas and caristas are made by wrapping a length of woven fabric around a separate bottom piece, with added handles. Promoters for Proyecto Zacapa Chiquimula (PROZACHI) introduced these

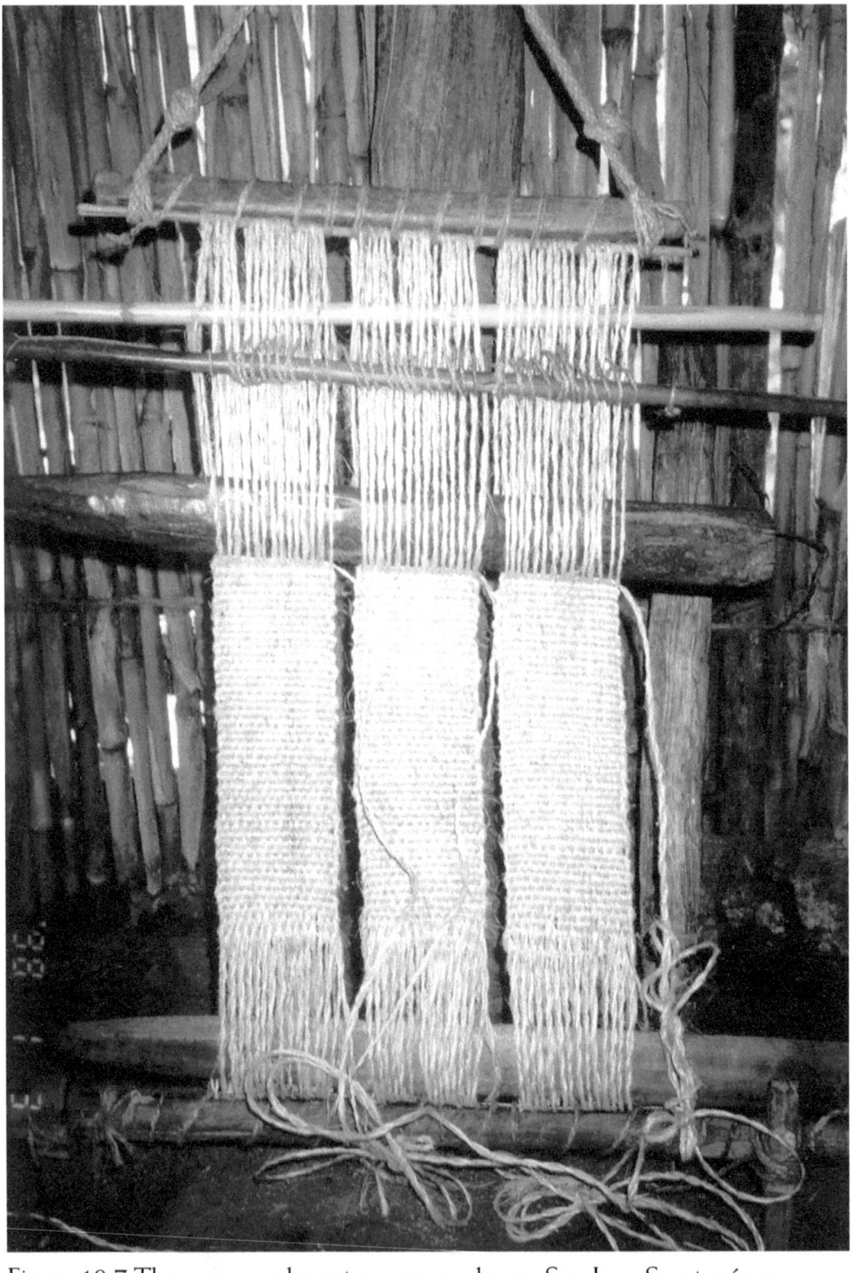

Figure 10.7 Three mecapales set up on one loom, San Juan Sacatepéquez, Guatemala.

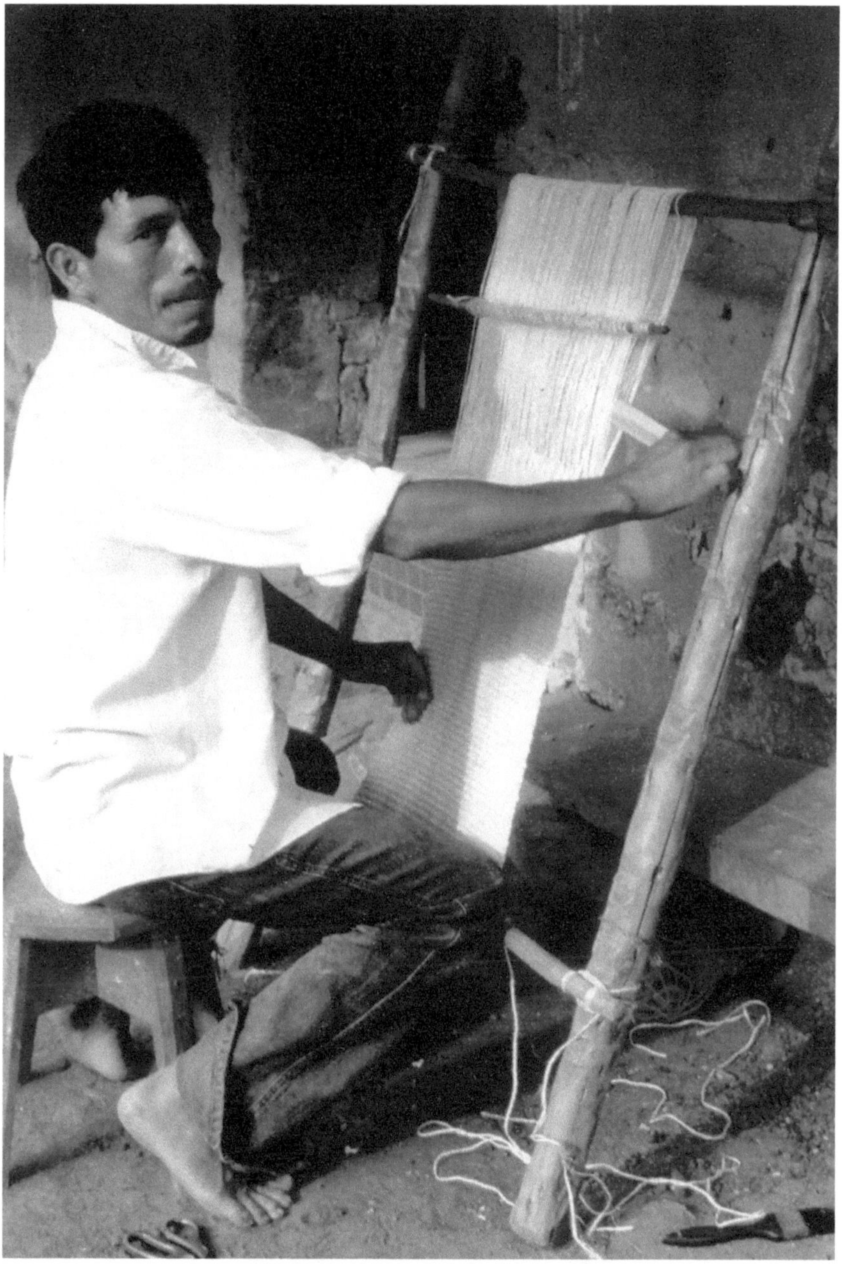

Figure 10.8 Candelario weaving a bag on a vertical-frame loom, Comapa, Jutiapa.

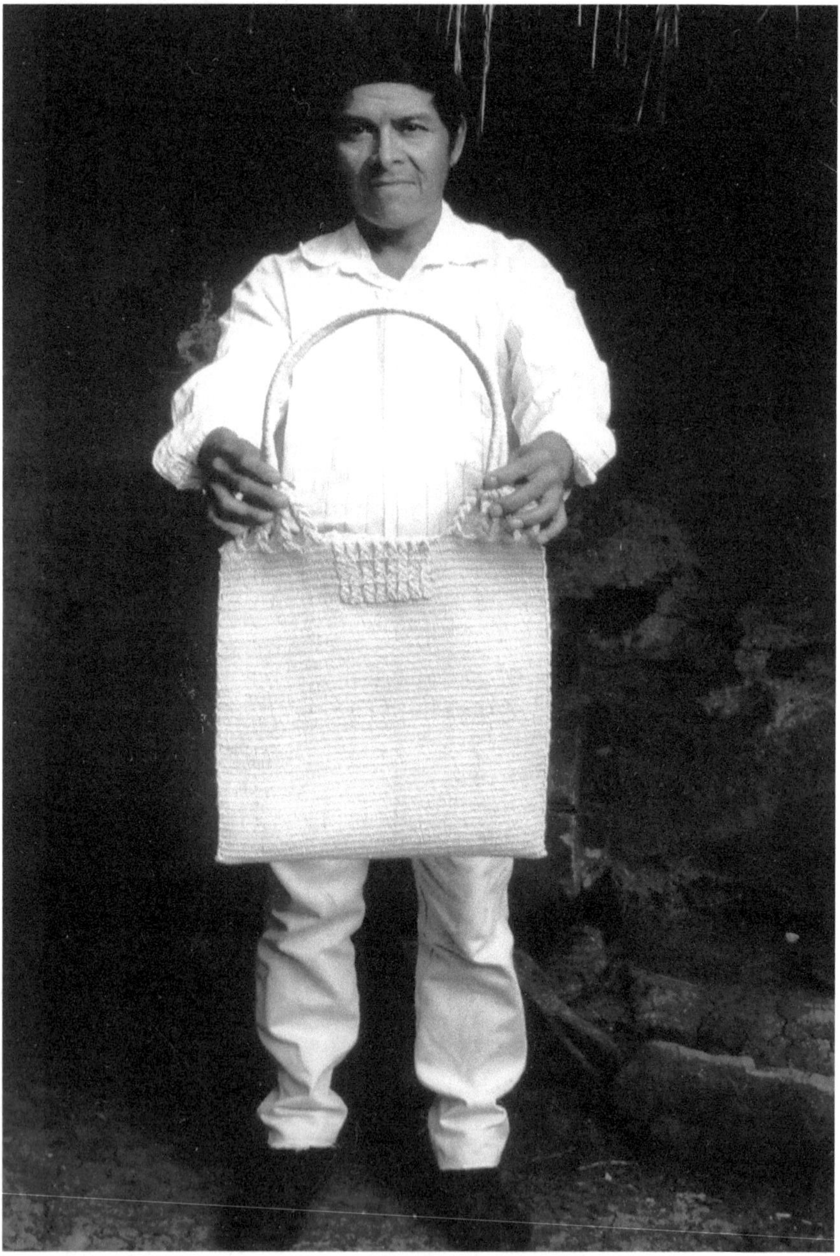

Figure 10.9 Toribio with a fancy bag that he made, Comapa, Jutiapa.

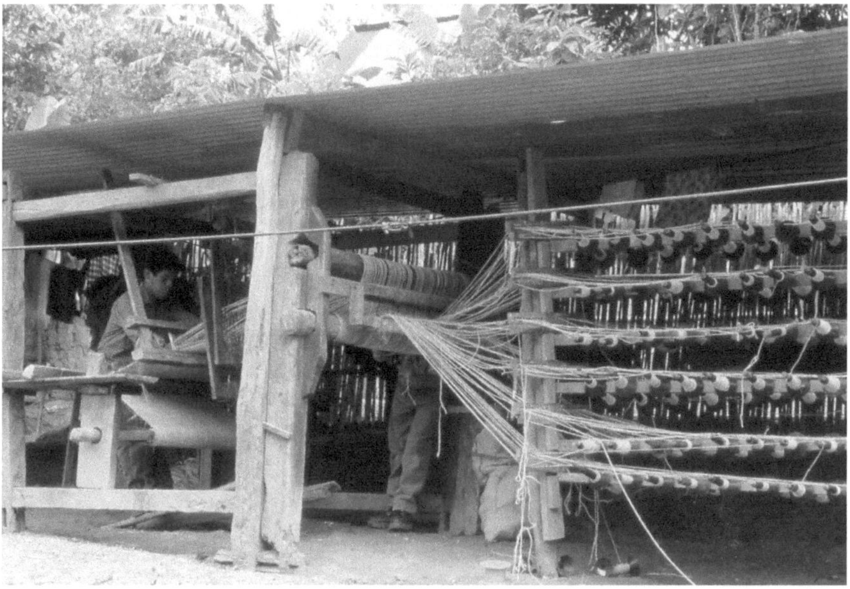

Figure 10.10 Alfredo weaving on a treadle loom, San Cristóbal, Alta Verapaz.

designs in the 1990s, and these were the bags that I saw for sale in the Peace Corps office.

Treadle Loom

Treadle looms, also known as shaft looms, have foot pedals that raise and lower the sheds (Seiler-Baldinger 1994, 82). Large items such as sacks, rugs, and fabric are woven on them. The treadle loom I saw had a continuous warp, in which tightly spun warp could be added indefinitely to one of 134 spools located behind the loom (fig. 10.10).

All square loom-woven single bags are constructed in the same way. The woven cloth is folded in half and the sides are sewn together with a piece of maguey thread to form the body of the bag (fig. 10.11). A handle is fabricated by wrapping the warp ends or adding a piece of rope.

Saddlebags or double bags are woven as two single bags, with the two bags made separately at opposite ends of the warp. After the first bag is complete the entire loom is turned around, or begun at the opposite side, and the other side is woven. The unwoven warp between the bags is braided toward the center, and a single strand interlocks them to prevent unbraiding (Collingwood, 1974, appendix 2) (see the earlier subhead "Sprang" in this chapter). For fancy saddlebags, the short section of unwoven warp on the ends that were attached to the loom is separated into groupings and wrapped with a strand of maguey using half hitches. This forms rings and

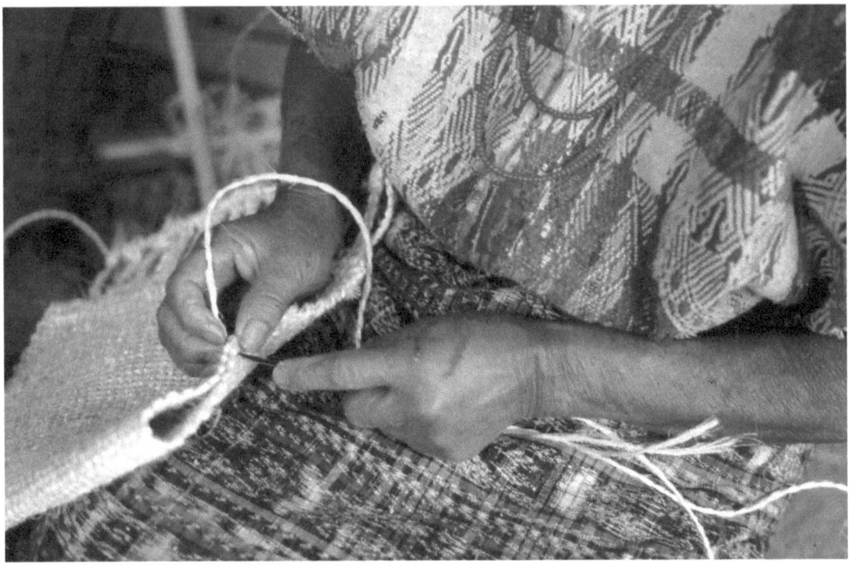

Figure 10.11 Basilia sewing the sides of a sembrador, San Juan Sacatepéques, Guatemala.

reinforces the ends (O'Neale 1945, 182). A drawstring is put through the loops (see appendix 3). On simple bags rings are not made and the drawstring goes through the loose warp ends.

In El Salvador, saddlebags were woven on unique looms; in one area they were tied with a rope to a tree branch, and the bottom of the loom was weighted down with stones or a heavy stick. There was such a peculiar sound when the weaving was beat that they became known as *chillote* (noisy) looms. In another area, two upright, planted poles at the distance of the saddlebag width were used, much like the Guatemalan hammock looms (Osborne 1956, 86).

Figure 10.12 diagrammatically illustrates many of the construction techniques discussed in this chapter.

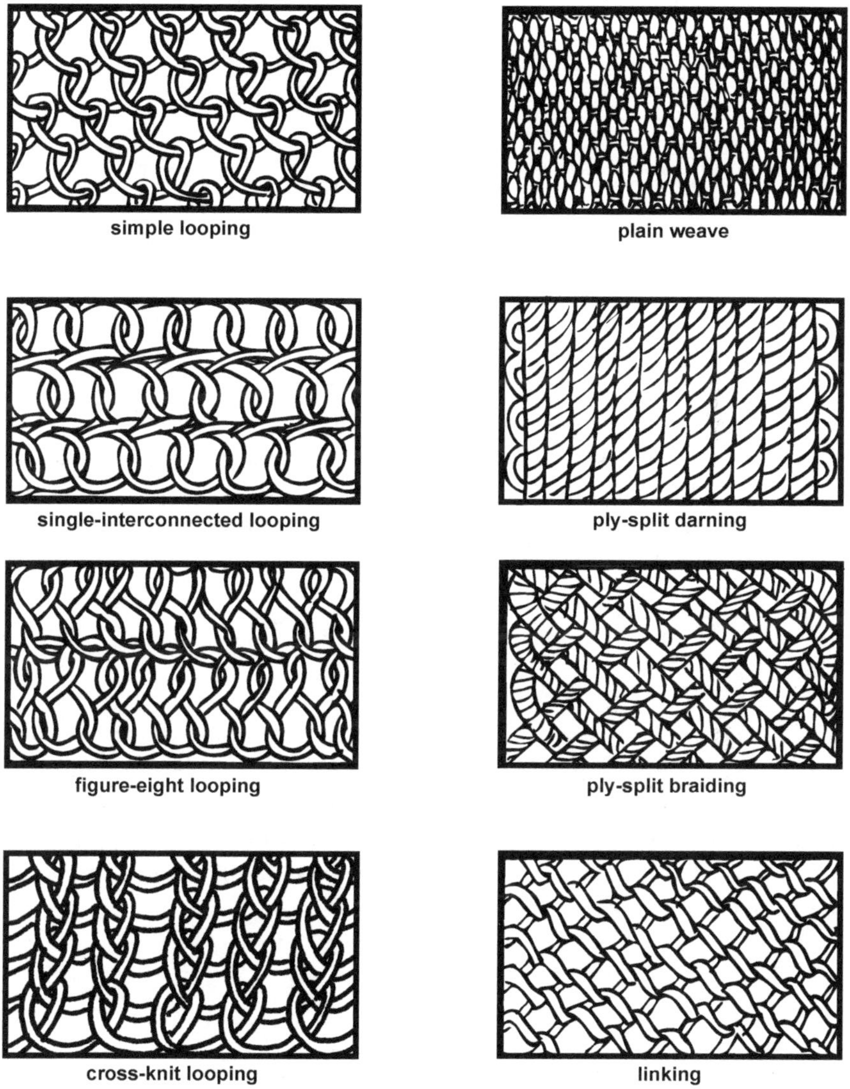

Figure 10.12 Technique diagrams.

11

Maguey Products and Their Uses

Lazo

Lazo or soga (rope) comes in all lengths and thicknesses. *Cordeles* or *gusano de pino* (pine worm) is a loosely spun length of maguey with inserted pine needles, used for decoration. Maguey rope is used for many things, such as being woven into beds (Wisdom 1940, 134) and, when finely spun, as strands for necklaces (Wisdom 1940, 37).

Red

The *red* is a round "cargo net" made to hold medium-sized objects and is the most common looped bag found nationwide. Redes can be used alone or around a large basket to contain the contents (photo essay fig. 11). Redes are made in different sizes with a variety of loop sizes depending on the intended use. For example, a *red* made for holding pottery has bigger loops than one for mangoes.

Small Bags

Small bags are a standard man's accessory, because men's traje does not have pockets. Objects inside of the bag are often wrapped in cloth. While this important accessory is usually associated with men, in a few areas women use them, too, although women mostly transport items wrapped in a woven cloth that is placed on top of their heads. As with other aspects of traje, specific bag styles, sizes, materials, techniques, colors, and patterns can be traced to different municipalities (see appendix 1). Household items are often stored in small bags, which hang from rafters and nails in many homes. The following are the different varieties found in Guatemala.

Morral

The morral is a small, square, plain or striped, tightly looped bag. In Concepción Tutuapa this bag is called a *sivadera*. It is usually worn with the strap over one shoulder or across the chest.

Bolsa

The bolsa is loom-woven. *Bolsa de mania* is the eastern term for these bags. Special loom-woven bags made in Olopa are *glincas* (round bags) and *caristas* (rectangular bags). *Sembrador* is a term used in San Juan Sacatepéquez for a small loom-woven bag for holding corn kernels during planting (photo essay fig. 12).

Matate

The matate is an open, stretchy, square, loose, looped bag. It is worn with the shoulder strap across the forehead, much like a mecapal, with the bag resting on the person's back. Mostly worn by men, O'Neale observed women in Cobán wearing bags in this style (1945, 182). *Cachelpa* is the term used in Concepción Tutuapa for matate. In San Andrés Huista, matates are linked.

Alforja

The alforja, or saddlebag (also *arganilla* [eastern term] and *argana*), is sometimes called a double bag. Alforjas are used on a horse or worn over the shoulder with one bag in front and the other in back. Saddlebags are always loom-woven.

Costal

Costales are large sacks made to hold small to medium loose items such as coffee or corn; they are woven on treadle looms.

Mecapal

The mecapal (tumpline) is a band that is three inches wide and twelve inches long. It is placed across the forehead and a rope attached to both sides of the mecapal is tied to cargo, which rests on the person's back. Men are able to carry very heavy loads using mecapales. Osborne called them "an article of dress and a man's prized possession" (1965, 134). Reina stated that they are "a symbol of strength and economic reliability to the Poqomám" (1969, 111). In the words of Feldman from his book *A Tumpline Economy*, "the load of trade rested on the shoulders of the porter, which is how items were transported on well-defined trade routes" (1985, 13–15). The same item is used for the strap that goes around the weaver's back for backstrap weaving.

Horse and Mule Gear

This gear includes *gamarrónes* (eastern term for halters) or *jáquimas* (western term for halters), *tenedores* (cruppers), *cintas* (cinches), and *enjalmas* (pads) (fig. 11.1). Gamarrónes and tenedores are made using the technique of ply-split darning, whereas cintas and enjalmas are loom-woven. O'Neale shows an example of a cinta constructed with spun maguey strands held in place by twined colored threads (1945, fig. 61); I did not see these.

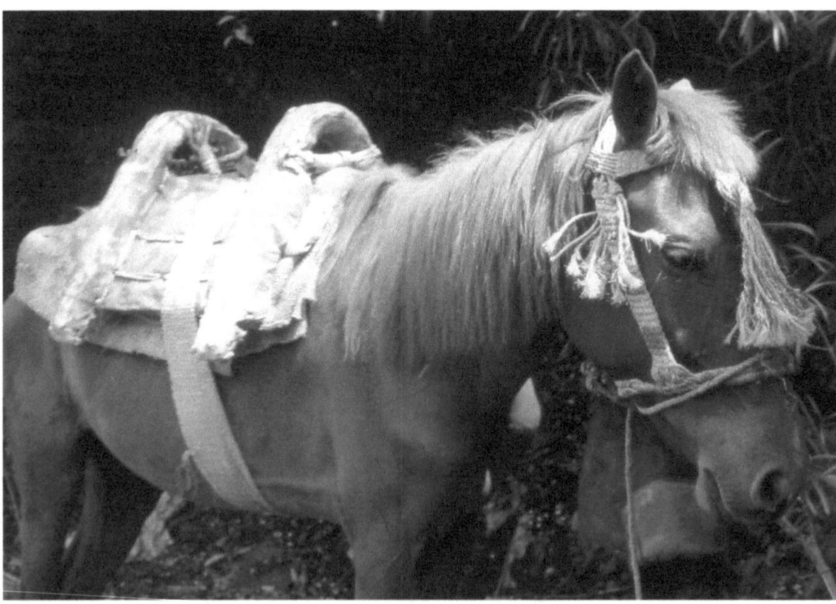

Figure 11.1 Mule with cinta and gamarrón, Concepción Tutuapa, San Marcos.

Bombas

Bombas (fireworks or other pyrotechnics) are very popular and were introduced to Guatemala by the Spaniards (Rodríguez Rouanet 1990, 238). Maguey is the preferred material because of its excellent compression characteristics (see the subhead "Wrapping" in chapter 10 for a description of a bomba's fabrication). Unlit bombas are placed in a tall metal tube with the wick sticking out; after the wick is lit, the bomba shoots up in the air and explodes. This common practice takes place during religious processions and was explained to me as a way to wake up the gods or to scare away evil spirits. In the 1930s bombas were set off in Chichi to let neighbors know when one returned home from a journey (Kelsey and Osborne 1961, 263).

Hamacas

Hamacas (hammocks) come in many sizes and are made with the technique of linking (photo essay fig. 13). Where it is hot, they are in front of every house, and small hammocks, where babies are rocked to sleep, often hang in several rooms.

Miscellaneous Products

Coiled baskets, placemats, doormats, decorative figures, animal muzzles, livestock whips, furniture, shoes, and many woven decorative items are made from maguey, as are coronas (crowns) and colorful strands used to decorate hats and vehicles. When visitors are in Esquipulas on a religious pilgrimage, they often buy palm hats with maguey strand decorations or long strands that they attach to their vehicles. In this way, everyone who sees the hat-wearing person or their vehicle knows that they have been to Esquipulas.

Crosses or arches placed in front of churches are also made from maguey (García-Mendoza 2006, 184). Wigs made from maguey fibers are worn with dance costumes (Kelsey and Osborne 1961, 104). Maguey paper is made by boiling small pieces of pita (spun maguey fiber) for three hours in water and caustic soda, which breaks down the fibers. The softened pita is washed, ground, soaked in water, bleached (to whiten) overnight, blended with water, and finally placed on a screen to dry.

Part III

The Gift of Life

12

Maguey Economics

Besides being utilitarian, the primary role of maguey products is economic. This is not new, and when I inquired about maguey's origin, many people told me that the land "gave" their grandparents maguey as a way for them to make money for survival if crops failed or there was insufficient agricultural land. In the words of a woman from San Pablo la Laguna: "La semilla del maguey nos concedió el Creador. Es un regalo que El nos dio y se puede sembrar en qualquier clase de tierra, sin importar la calidad del suelo, es un bendición" (The maguey seed was given to us by the Creator. It is a gift from our God, which can be planted in any class of land, without importance of soil quality, and it is a blessing) (Berger de White and Ivic de Monterroso 2008, 136). Others told me that "maguey is a symbol of life" because money can be made from selling maguey products, and therefore "gives life." In addition to growing just about anywhere, an important advantage of maguey over many other crops is that it can be harvested at any time of the year, thus providing a year-round income.

The products' journey from maker to market begins with location, and maguey items are sold wherever there might be a potential customer. This includes markets, festivals, tourist sites, buses, and busy street corners. Weekly markets occur everywhere and have been a part of Guatemalan society for centuries, and regional communities are linked in a rotating schedule of staggered market days, or "solar system" (Nash 1967, 87). In this way vendors can sell their products in many nearby locations every week.

Generally, "traditional" products are sold in regional markets, and "contemporary" items are found in gift shops. Tourism does not seem to have affected the maguey industry very much, and most visitors buy brightly colored woven pieces, often unaware of the maguey articles in the markets.

The weekly Concepción Tutuapa market is one of the biggest that I attended, and on Thursdays the normally quiet, empty, dusty town is transformed when it fills up with vendors selling their wares on the streets. A steady stream of people enter the community from the outer edges, and because of the early-morning chill, they are usually bundled in towels and blankets. Many lead cargo-laden mules, which are tethered in nearby fields after being unloaded. I always locate the mules so as to admire the maguey fittings that adorn their bodies and to take photos, much to the amusement of bystanders.

The maguey arena is at a nondescript intersection, past the clothing, produce, steaming tamales, and baskets of *xeca* (wheat bread). The bargaining starts around four in the morning, when men and women from the outlying aldeas begin arriving by bus, by horse, or on foot. The buyers are already waiting, and as products are revealed from within wrapped bundles, the frenzy begins. Bargaining can be quite intense, with strong words spoken on both sides about the quality, price, and financial needs of each person. It always amazes me to watch a man successfully bargain a woman down to a price of about twenty quetzales for a beautiful bag that took her at least a month to make.

The accepted protocol is that, when a seller reveals his or her wares, a buyer will grab an item and begin haggling, usually starting at an amount much lower than the asking price. If the seller does not come down enough, the buyer will let go and another will take his or her place. The understood code of ethics is that an object is not available for bidding until only the seller is left holding it. The action continues until there is agreement on the price and the buyer adds it to his or her pile of purchases or the potential buyer lets it go and a new buyer moves in. If the product is inferior or the price is too high, sellers may stand on the corner for an hour or more until they give up, rewrap the item for another time or place, and merge back into the general flow of the market.

In the beginning, I bought lots of maguey without really knowing what I was doing, and bystanders would come up to me after a purchase and ask why I bought the fiber that I had, as they could tell from a distance that it wasn't very good. After a bit more experience and advice from the locals, I eventually learned the difference and began to notice that when a seller with nice soft fiber entered the selling arena, he or she usually sold out within minutes.

Fibers sell by the manojo (usually about ten to fifteen leaves in this region) or pound, whereas sivaderas, cachelpas, and enjalmas sell individually. Strands of lazo, mecapales, and cintas sell by the dozen. Redes sell by the pair, I believe because they are used to hold cargo on both sides of a mule (fig. 12.1). Synthetic materials for rope from nearby Sipacapa and

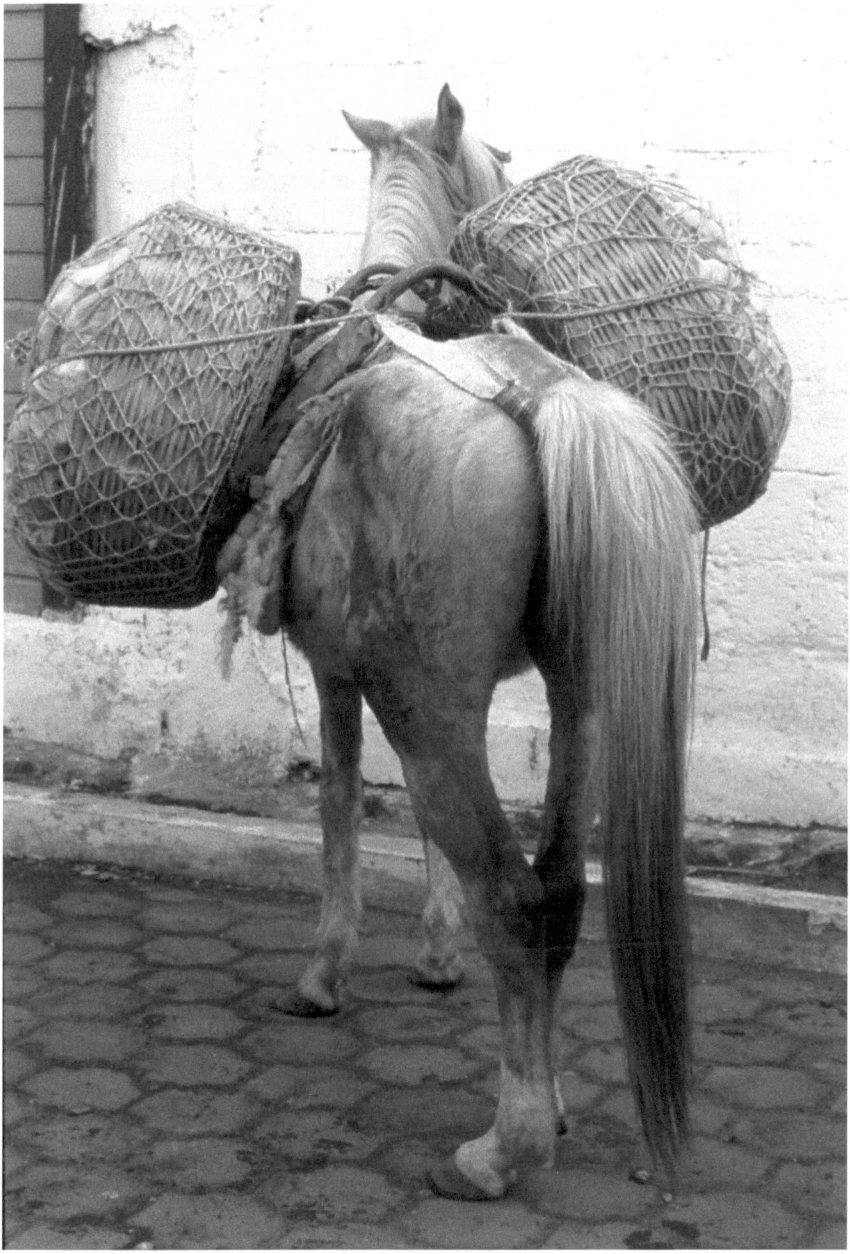

Figure 12.1 Mule with two redes and a tenedor, Concepción Tutuapa, San Marcos.

synthetic morrales from Comitancillo are sold by middlemen along a sidewalk near the entrance of town and not in the maguey sector. Occasionally a locally made, crocheted, recycled plastic shopping bag morral appears in the maguey area.

Next to the maguey products, wood-fired pottery is sold, and this gives me the opportunity to talk to pottery vendors as well. As it turns out, many of the same women make both pottery and maguey items, but currently pottery economics is better, so more women are creating clay pots than maguey articles.

Many of the buyers resell the items at local markets in Mexico and Guatemala City, and several go to the department of Huehuetenango, where the majority of men and women still use maguey morrales. Large quantities of fiber and redes are bought for resale to coastal farm and ranch workers, and Sipacapa rope makers also buy fiber. Many fiber buyers are from the lower-elevation aldeas of Yamoj, la Laguna, Sochel, los Encuentros, and Sichivila, known as la Vega, where mecate, a high-elevation maguey species, does not grow. The inhabitants of these areas are considered the best morral makers in the region. The maguey market is usually over by 8 a.m., and those who sold products now have money to spend elsewhere.

Compared to Concepción Tutuapa, the Friday and Saturday markets in San Juan Cotzal are very small. The void is filled by many shops, which are large and full of merchandise. However, unspun fiber and redes are sold only during the weekly market on the street or by the concrete stairs near the plaza. In 2001, there were also two hundred women working with Proyecto Quiché and thirty women with Asociación de Desarrollo Integral Tetz K'usal creating maguey placemats and netted maguey ceramic pot covers. When I paid a visit in 2002, they were looking for markets outside of the community, and when I returned in 2008, they were no longer working on these items because a substantial market was never found.

Since there is not a weekly market in San Pablo or San Marcos la Laguna, maguey products are sold to neighbors, sold in a few small tourist shops, or taken to markets in Panajachel, Santiago Atitlán, and Sololá. Usually each person has a regular market and location where they sell, such as is the case with Pablo, the rope maker from San Pablo (see chapter 3).

Every Friday, Pablo travels in the back of a pickup truck from San Pablo to San Pedro la Laguna, which involves several transfers on a bumpy, dusty road. He then embarks on a half-hour boat journey to Santiago Atitlán and climbs up a steep hill to where he stands in the same spot each week (fig. 12.2). Sometimes orders are placed in the market to be picked up at a later date, such as the one-pound balls of thick single-ply strands Pablo was

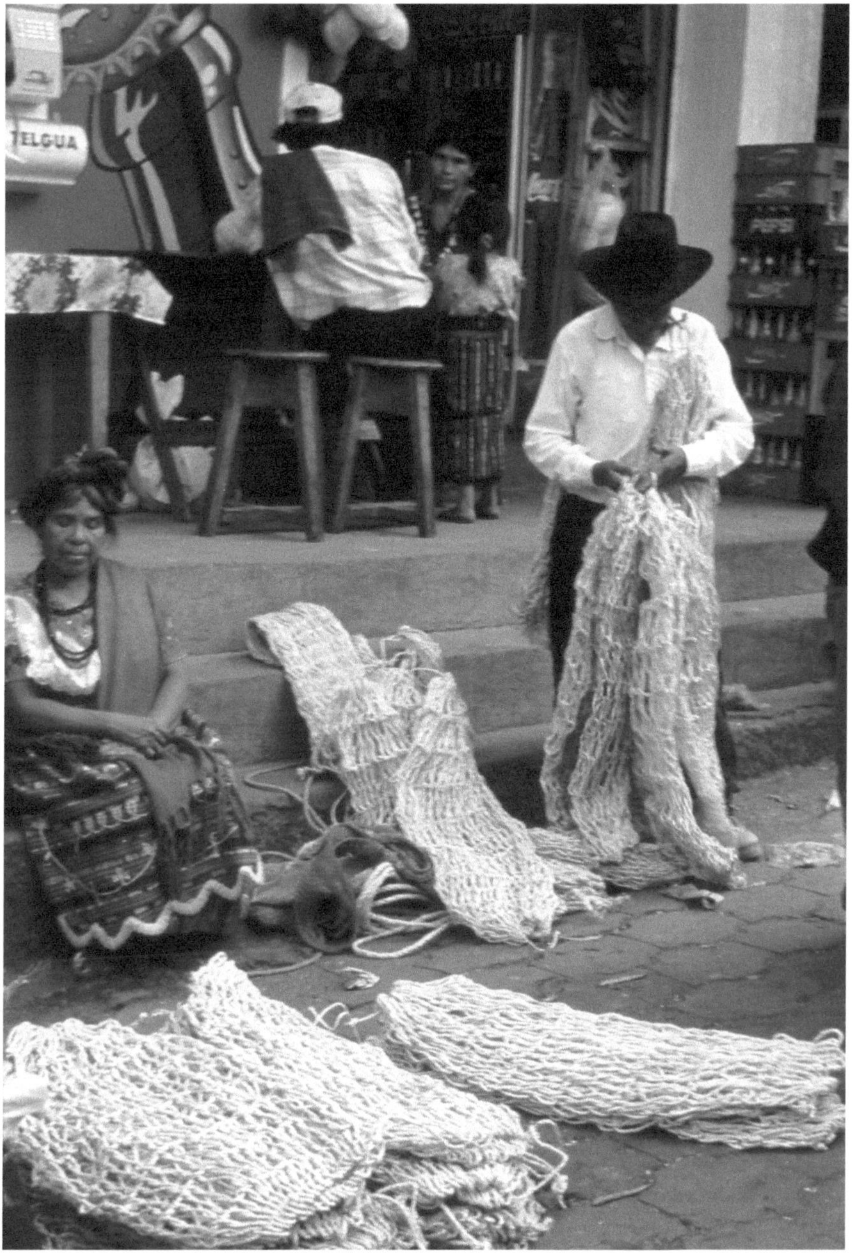

Figure 12.2 Pablo and María selling redes in the market, Santiago Atitlán, Sololá.

spinning in 2007, when I paid a visit. Each ball took him about half an hour to spin and he told me that he would make about four quetzales apiece for his efforts.

In 2009 I ran into María, Pablo's wife, wandering around the Santiago market with several redes draped over her shoulders. She told me that Pablo gave up his sidewalk space the year before, and now they take turns roaming through the crowded streets to sell their wares.

The inhabitants of San Juan Sacatepéquez aldeas usually take maguey products to the central market in Guatemala City on Friday or, as happened when I was visiting Pablo's family, a buyer will come by the merchant's house. Pablo's customer picked up three dozen sembradores for fifty quetzales and two dozen mecapales at twenty quetzales.

Comapa has a small weekly market, but maguey products are not sold there. Most morrales, redes, and gamarrónes are bought directly from the makers by middlemen who resell them in Jutiapa, on the Pacific coast, El Salvador, or Guatemala City. The few bolsas, alforjas, and hammocks that are sold in Comapa are crammed into a few small stores that line the streets. Beside the local products are tenedores from San Pedro Pinula, lazo from Jocotán, and sembradores from San Juan Sacatepéquez. Sometimes fiber is sold as well.

Jocotán's maguey products are sold in the Sunday market, where brightly dressed Ch'orti' women come down from the thirty-four aldeas to sell their wares. The maguey sector is located in the street near the bus "terminal." Here buyers and sellers converge in the early hours and piles of hammocks, redes, and lazo line the streets (photo essay fig. 14). Often women with rolls of palm mats and bunches of brooms balanced on their heads walk by on their way to the palm market. By 2009 most of the hammocks and redes sold in the market were plastic, not maguey.

Also in Jocotán, the groups Coordinación para el Desarrollo Integral Maya Ch'orti', Asociación Solidario Desarrollo Integral, and Proyecto Coop Español (Desarrollo Regional Municipal y Humano en el Área Ch'orti' de Guatemala) are working with maguey weavers to produce items for the international market. Another organization, Asociación Artesanal Ajpat'nar Ch'orti', collaborates with maguey, cane, and palm artists to market contemporary products as well. Aniline-dyed materials and product design expertise are provided, and the work is completed in the artist's home. In 2007, two students from the University of Michigan developed maguey lampshade and chair projects for the company, and that same year a representative went to a Latin American trade show in Los Angeles to market them.

Olopa has a very small weekly market but maguey items are not sold there. Instead, the office of AMO (founded in 1971) has a shop where glincas and caristas as well as *bolsas de mano* (handbags), *bolsas de hombre* (men's bags), *bolsas infantiles* (children's bags), *juego de individuales* (individual placemats), *cinchos* (belts), *zapatos* (shoes), and *muebles* (furniture) are sold. I spoke with the employees and they told me that the then-mayor's wife was instrumental in much of the group's current progress and success. The group is looking for new markets, and a representative went to Exporural, a trade fair in Chile; to Copán, Honduras; and to the Latin American trade fair in Los Angeles to find them.

There are many places in Alta Verapaz where one may buy maguey items. In the San Cristóbal Verapaz market, one-pound bundles of unspun fibers are sold by thirty-five-year-old Estela (photo essay fig. 15). I observed her selling about fifty pounds in two hours. In 2002, she sold a pound for three quetzales and bought redes at seven quetzales a pair, to resell. By 2007, one pound of maguey sold for five quetzales and two redes had a price of about eight quetzales. Estela's family has been selling maguey in the market for forty years. May through August is when the largest quantity of maguey fibers is sold, and in January and February men buy redes by the truckload to resell in Guatemala City's main bus terminal. In the 1990s, one thousand quintals of maguey were sold per month in San Cristóbal Verapaz (Camposeco Mateo 1994, 48). Also in San Cristóbal, the gift shop in the Katinamit Museum, which is affiliated with the Centro Communitario Educativo Pokomchi', has maguey items for sale.

A few places in Cobán sell traditional weavings and handicrafts, including maguey products. They are Casa D' Acuna's gift shop, Xkape Kob'an (Café Cobán), and Aj K'uub' Anel' (Artisans Association of Verapaz), which collaborate with the Asociación de Caficultures de Tamahú y San Cristóbal, Alta Verapaz. Another organization supporting maguey artists is Artesanías de la Agexpor, and promoters work with women from the aldea of Vista Hermosa, San Cristóbal. They make maguey fiber baskets (Magaña 2001).

Attaining the "gift of life" can be a challenge. Key is finding a sustainable market where buyers appreciate handcrafted items made from natural materials, and whether an item is considered traditional, contemporary, utilitarian, decorative, craft, or art is often in the eye of the consumer. Currently the products of Concepción Tutuapa, Jocotán, Olopa, Cobán, and Comapa are sold farthest from their production centers, and there is a great likelihood that maguey items will be found even farther from their points of origin as long as the demand is there and Guatemala continues to open up to a global economy.

13

Maguey in Transition

As in many places around the world, early inhabitants worked with what they had in their immediate surroundings to create items necessary for their daily lives. When the possibilities of the fibrous maguey plant were discovered, the tradition began. As clothing was desired, maguey fiber was spun and woven into articles of dress, and when different-sized bags for containing and carrying a variety of items were required, maguey was found to be the perfect material for making them, probably because the fibers were long, strong, durable, pliable, and uniform (Schery 1972, 174).

Maguey fibers continued to be important during colonization, and when horses and mules were introduced, the fiber was again found to be perfect for making horse gear. Over time, missionaries, explorers, development workers, artists, and even tourists introduced new ideas, influencing the maguey industry. The maguey tradition evolves and expands as needs, values, fashion, and cultural migration dictate materials, techniques, and products.

In 1861 Squier put into words what many already knew: "The value of the agaves is enhanced by their indifference to soil, climate and season; by the simplicity of their cultivation, and, by the facility of extracting and preparing their products" (27). In the Yucatán, maguey was referred to as "green gold" in the 1930s because of its high binder-twine fiber value (White and Foscue 1939, 151). In central Mexico, the Otomi called it "Mother Magey," because everything necessary for survival in their harsh environment could be harvested from this plant (Granberg 1970, 18). The Nahuatl referred to *metl* as "the miracle plant" (Valenzuela-Zapata and Nabhan 2003, 5).

In Guatemala, most likely, at one time the cultivation and harvesting of maguey occurred in most of the countryside. By 1990, every department except for Quetzaltenango, Retalhuleu, Esquintla, and Izabal reported harvesting maguey (Domingo 1990, appendix 8). Camposeco Mateo listed the departmental maguey production centers as Alta Verapaz, Huehuetenango,

San Marcos, Jutiapa, Chiquimula, and Guatemala (1994, 47). When the civil war ended in 1996, accelerated changes came and opened wide the doors of globalization. Knapsacks and nylon bags are now more common than morrales, and by 2001, many maguey-producing areas began to shrink into isolated pockets scattered around the republic.

Besides new products, plastic fiber, introduced in the late 1990s, is rapidly replacing maguey. The brightly colored, unspun strands of plastic come on large spools and sell for eight quetzales a pound in the shops and markets. After the plastic reaches the aldeas, it is spun on the same wheels and made into the same products once made from maguey. Some people like working with plastic and others do not. Maguey fiber extraction is very labor intensive, smelly, and dirty, but plastic costs money. Market demands often determine which material is worked and for which product. In Concepción Tutuapa, maguey is preferred for cintas and mecapales because when these items are made from plastic, they tend to double over when used. In San Juan Cotzal plastic is now common because so many men have left town looking for work that there are not enough men left for the fiber-extraction gatherings.

A few people I talked to in Cotzal said that they preferred plastic morrales over maguey ones because when maguey gets wet it weighs more and rips easier than plastic. Others do not like working with plastic because it does not spin as tight as maguey fibers, whereas many like it better because it does not irritate the skin as much when leg spinning. Others find plastic preferable to maguey because the spun fibers do not have to be trimmed of their "hairs." Some people do not like plastic because it "burns" (deteriorates) in the sun. Most said that whatever the market wants, they will provide.

In some areas, decline of the local maguey industry is attributed to loss of plants to the *plaga* (plague). In Comapa an insect called *ronrón* (the common name for a beetle) destroyed many maguey crops. There are two different insects. The larger one is only a problem during the rainy season and eats the roots at night, burying deep in the soil during the daylight hours. The other, smaller one is the agave or black weevil (*Scyphophorus acupunctatus*), which eats the leaves year-round, eventually killing the plant (Aldana and Estrada, 2002, 392).

In a couple of areas, people told me that bags filled with an unknown powder were dropped by airplane and distributed from trucks in 2003. While it was thought they might have been part of a nationwide attempt to counteract fruit flies, about this time a plaga also hit, and this was either a coincidence or related to the mystery powder. Some believe that it was a conspiracy of the Guatemalan government against them, because this unknown substance "burned" many maguey plants and forced the use of plastic, fabricated in Guatemala City.

I also heard that the main reason for the decline of the maguey industry is that, fifteen years before my arrival, cheaper fertilizer became available, which gave people better and more abundant crops, thereby diminishing the need to make and sell maguey products.

The coffee industry has had both positive and negative influences on maguey. Native to the Arabian Peninsula, it was introduced to Guatemala in the 1860s as part of a crop diversification and economic development program when the market fell out of the natural dye industry. Since then, coffee has established itself as one of Guatemala's top exports.

After the coffee beans are picked, they must be stored in "breathable" bags, and maguey is the perfect medium for this. For many years maguey costales were manufactured in factories, which increased the demand for maguey. Today, however, jute (*Corchorus capsularis* L. [Tiliaceae family]) is the preferred costal material, which decreased the demand. Native to India, jute's fiber use has been known for a long time and is documented in early Sanskrit writings and the ancient Vedas (Standley and Steyermark 1952, 327). Like maguey, jute's fibers are extracted from the stems of the plant by retting (soaking in water for ten to fifteen days) and then removal from the soft leaf. The fibers are woven on floor looms into burlap and sewn into costales. Jute replaced maguey probably because it is cheaper to process, yet retains the "breathable" quality that coffee-bean sacks require.

Another factor leading to the decline of maguey cultivation is that many farmers replaced maguey plants with coffee trees when the value of coffee was high and they could potentially make more money selling coffee beans than items made from maguey. Even in locations where coffee does not grow, many people travel to coffee farms during harvesttime to earn cash and are therefore absent from the maguey scene.

In the department of Alta Verapaz, many workers mentioned cardamom (*Elettaria cardomomum*), not coffee, as the crop affecting the maguey industry. Native to India's Malabar Coast, cardamom was brought to Guatemala in 1917 by a German coffee grower. Related to ginger, the ground seeds are an important spice for classic Indian cuisine, and in the Middle East cardamom powder is mixed into a thick syrupy coffee drink called *gahwa*. Cardamom is so valuable that all other activities are dropped when it is time to harvest, dry, and sort the seeds.

Political, economic, and social trends also affect the maguey culture when much of the rural population travels away from home for education and work. Mass migration to the United States eliminates part of the local workforce and alters personal values, further impacting the maguey industry. Currently, in many areas, the youngest person working with maguey is in his or her seventies, and the children or grandchildren have little interest in continuing the tradition.

In contrast to the declines, maguey is being promoted in a few locations. One place is San Marcos la Laguna, where a young environmental education teacher working with Amigos de Lago Atitlán strives to inform the public about the importance of native plants, including maguey. He works in the schools, and with Mujeres Luna Kaquikel, a group of about thirty women. They have a nursery full of native species, including one thousand maguey plants that will eventually be transplanted around the lake (mostly as ornamentals, with one plot designated for "traditional maguey use"). Also in San Marcos, a group of young men called Hombres de Maguey is promoting the maguey tradition (Heisse 2000, 11). In Olopa, on land managed by AMO, maguey is being planted as well.

Highlighting maguey as a tourist attraction is an idea being promoted near San Cristóbal Verapaz in the aldea of Pamuc by a group called Hostal Nuboso Maya-Pokomchi Pamuquense. The members state in their brochure that, "besides learning how to make tortillas, maguey twine making can be learned." Basic accommodations and food are provided, as well as guided tours into a nearby cloud forest (Centro Communitario Educativo Pokomchi' 2001).

There is also a small developing movement of artists in several communities who produce and sell "contemporary" maguey products such as rugs, baskets, and decorative figures. One of these is the newly formed women's group in San Pablo la Laguna called Mujeres de la Associación Q'apoj Saqkiij (Berger de White and Ivic de Monterroso 2008, 179–182).

In Santiago Atitlán, approximately eighty textile workers make products for the Cojolya Association of Mayan Women Weavers, a nonprofit organization founded in 1983. Their office, Weaving Museum, and a gallery shop that opened in 2002 are located on the main road between the boat dock and town. The group has a Web site where one of the bags advertised is "Espadrille," made from backstrap-woven cotton thread and "very fine, hand-spun maguey twine." It is described as "a unique handbag . . . designed to sustain the traditional Mayan rope-making craft of maguey" (Cojolya Association of Mayan Women Weavers 2010). The organization also offers weaving instruction for individuals or groups, demonstrations by weavers in their homes, and an opportunity for visitors to see the maguey plant and the process of making twine.

When a new mayor took office in Olopa, the Oficina Municipal de la Mujer was formed and AMO lost its technicians and central office location. Even so, AMO still has the maguey monopoly in town, and its new office on the outskirts is full of interesting maguey items, many of which are shown in their brochure and new color catalog (Asociación de Mujeres Olopenses 2004, 2009). In 2009, the cooperative Amor Ch'orti' was formed and includes maguey, palm, and cane artists. It is affiliated with Acción Contra el Hambre, Mancomunidad Copanch'orti', and Gobierno de Navarra (Artesania Ch'orti' 2010). They opened a small shop at the entrance of Jocotán in 2010.

The only mechanized process I saw in Guatemala was an electric rope spinner in Santa Cruz Verapaz. Apparently a fiber-extracting machine once existed in Camotán, Jocotán's neighboring municipality. The local Belgian missionaries initiated the project, but no one maintained the machine and, by the time I visited, it was long gone. In Comapa, I heard rumors that a motorized fiber-extraction machine was to be installed, but, in 2007, I learned that this was a project of the departed mayor and, when he didn't win reelection, the idea was conveniently forgotten.

Transitions are ongoing. Even in the relatively short time that I researched maguey textiles I saw many changes, and though I categorized techniques, processes, and products, there is a possibility that I missed something in a remote location. Custom, fashion, need, and market demand determine production, and transitions will likely continue as long as maguey grows in the countryside.

14

Conclusion

Maguey is possibly the oldest continuously used plant fiber in the region. Osborne writes: "The spinning of fibers from vines and fibrous leaves was practiced long before cotton was woven into textiles" (1956, 81). When enhanced by its value as a historical record and a visual language (see appendix 1), maguey is very important. However, globalization and modernization are affecting maguey's use, making its future unclear.

As late as the 1980s, maguey morrales were indispensable accessories and everyone had at least one. During the civil war, *campesinos* (peasants) were advised to use morrales so that soldiers could see what they were carrying. During this time, there was less movement of people and goods and greater reliance on local materials, which assured the craft's survival.

When the civil war ended, the country opened up, and an influx of outside products such as nylon knapsacks and plastic fiber began replacing maguey. The Central American Free Trade Agreement (CAFTA), signed in 2004, could have further impacts on the maguey industry, as it will allow for additional imported goods.

Besides new products and materials, if maguey comes to be viewed as having a low economic and social value, the industry could suffer. When young adults move from the social environment of a subsistence to a cash economy, the making and use of maguey utilitarian objects might be seen as old-fashioned, as undesirable, and even as a symbol of the country life that many are trying to "escape." In quite a few communities often you will find only one or two elders who continue to work with maguey.

Fortunately, when I began seeking maguey workers and their products, I found that they were still around, only less visible. A glimpse into the workings of the existing industry reveals some interesting aspects. One is that the lives that revolve around maguey in many ways probably mirror those of their ancestors and can provide historical insights, as well as showing us the changes taking place.

Another aspect of the existing industry is tool, technique, and product specialization. Obviously, components of fiber extraction are based on climate and species, which also affect fiber softness, how it is spun, and possibly what is made. Looping and leg spinning go hand in hand, whereas spinning wheels are found where techniques such as loom weaving and linking are used, which have larger working spaces.

Presumably, due to regional isolation, methods developed independently or were introduced as discovery and need dictated. Maguey knowledge and skills can also travel via cultural migration—for instance, when the ancestors of those now living in San Andrés Huista and Comapa emigrated to Guatemala, they brought their maguey skills with them. There could be processes that I did not see, and crossovers between textile traditions (such as crocheters in Olopa, who are now working with maguey) are inevitable. Even so, many people are resistant to change, and when I tried to get some morral makers to try new techniques, most were uninterested. Osborne observed this as well: "The Indian and his craft are bound together by the bonds of custom and tradition" (1965, 8). Another example is seen in the bag styles and techniques of Gustavus A. Eisen's 1902 collection, which are identical to those of today.

In contrast, a few people wanted to learn new techniques, such as a man in Comapa to whom I taught the cross-knit looping technique that I learned in Cotzal. He shared the information with family members, and each adapted it to their own style. When I returned eight months later, the man whom I had originally taught had made three maguey morrales, his son a synthetic morral, and his daughter-in-law a woolen one, all from the cross-knit technique (fig. 14.1). When I saw this, I understood how ideas move from one region to another and found it interesting to see which materials were used based on gender and age.

Tools can also provide insight, especially the looms. Where maguey is woven on backstrap looms, it is mostly for small, narrow items such as mecapales, cintas, and bag straps. In regions where backstrap weaving does not exist, only frame-type looms are found. One possibility is that frame looms, or the "false looms" Wisdom (1940, 161) saw, were initially used for weaving hammocks (similar to the South American looms), also a common item made in the East. From hammocks, the looms could have then been adopted for other maguey products, especially saddlebags, which would be difficult to make by looping or on a backstrap loom. This is also evident from the Salvadoran saddlebag loom, which is identical to a hammock

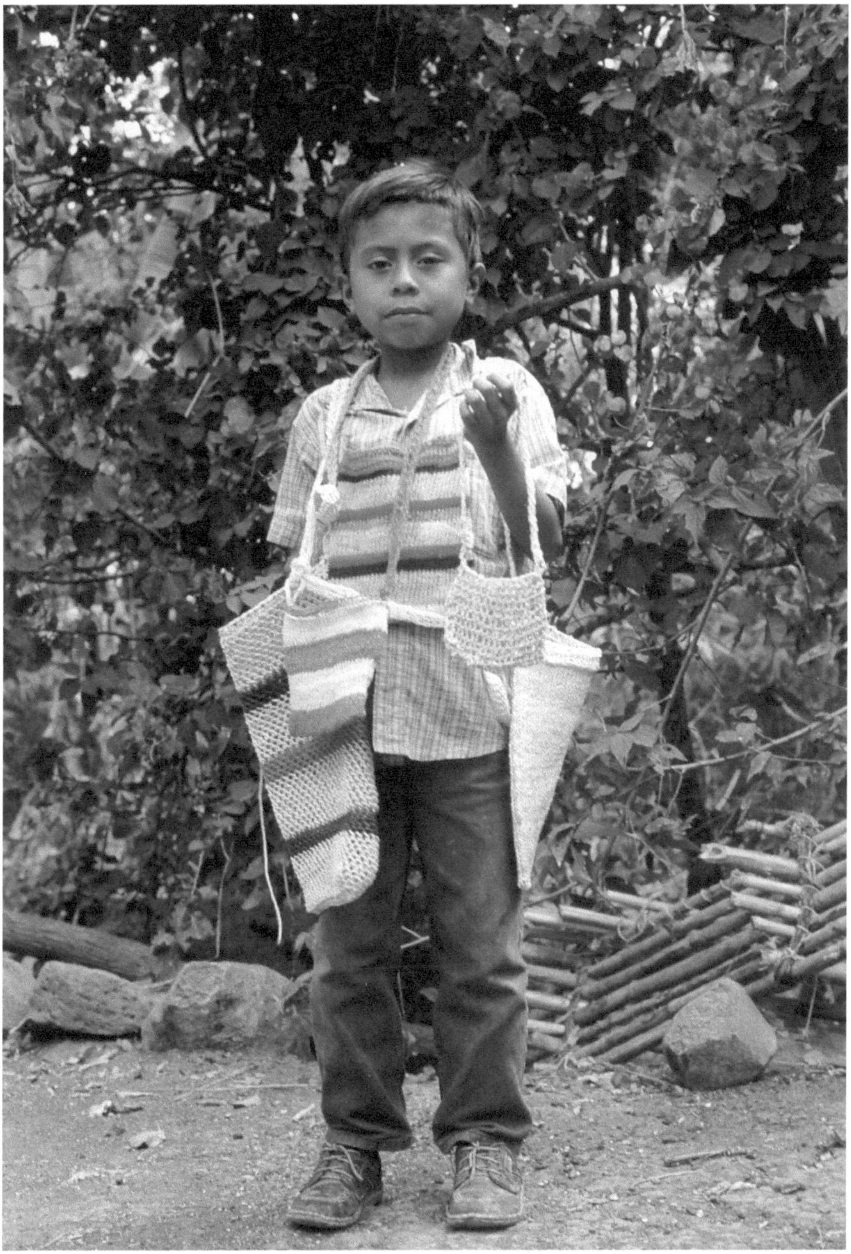

Figure 14.1 Alonso with the net bags made by his family, some using the cross-knit looping technique, Comapa, Jutiapa.

loom (Osborne 1956, 86). Since the Spaniards settled earlier in the East, and introduced horses, it is logical that the idea for the tools needed to create horse gear would have been introduced by them. The design probably spread, and while O'Neale observed maguey being woven on vertical-frame looms in the Occidente (1945, 35), I saw them only in the Oriente.

I never learned specifically why these looms exist, and weavers always answered me with the standard *"es la costumbre"* (it is the custom), meaning it has always been done that way and no one questions it. For most people, the treadle loom is cost prohibitive and so is found in only a few locations.

One aspect that I had hoped to find was the symbolism of maguey and its products. I asked many elders, for most likely this information was passed down in the oral tradition of their ancestors, but I gained very little from this approach and so I relied on my own interpretations. An artist in San Juan Cotzal told me that "morrales are important because they unite all of the indigenous people together, from the East, West, North, and South." A man in Comapa said that his beginning loops were equivalent to "children." That maguey is viewed as "sacred" also highlights the importance of this amazing plant, and gave me food for thought. I also understood its importance in other ways.

Maguey as a "symbol of life," a "life-giving" force, and a "gift" from the land deserves recognition, and maguey can possibly be called the "plant of life," for it can provide everything necessary for survival, even in harsh environments, such as where the Otomi live, who called it "Mother Maguey" (Granberg 1970, 8). In addition to economic and utilitarian benefits, maguey provides social, cultural, and spiritual outlets. The fiber-extracting work parties of San Juan Cotzal bring men together, and the market gatherings of sellers and buyers provide financial and social exchanges for men and women. The requirement that a young woman in Concepción Tutuapa must possess the skills necessary to make a morral before marriage illustrates its social importance, while the "queens of maguey" at fiestas and the "welcome to the land of maguey" and "comedor maguey" signs make clear its cultural significance.

Government-sponsored craft competitions, such as the one in which Santós entered a hammock and won first place for (see chapter 5), also give value to maguey. The inclusion of maguey in religious ceremonies in the form of bombas, coronas, and decorations shows maguey's ritual side, and the fact that "traditional" morrales are sold in the inner sanctum of Chichicastenango's market, where mostly locals shop, illustrates that tourism has not yet affected their production and that they are the "real thing."

As with many aspects of rural Guatemalan life, including maguey, it is difficult to predict the future. If petroleum products become less desirable, maguey could regain its importance as a natural, biodegradable fiber, which might bring about increased protection from plague, higher fiber-yielding plant varieties, mechanized extraction and spinning possibilities, plus new product ideas. As unforeseen challenges, influences, and markets affect Guatemala's maguey transformation, I am hoping the strand that connects the past with the present will continue into the future and, like a giant *red,* hold those who live in this "land of maguey" together, as it has for generations.

Appendix 1

The Language of the Bag

Guatemalan traje, including small bags, is distinct to specific municipalities and provides a visual identity to those who know the language. Distinctive bag traits are size, shape, design, color, material, and technique. While many are made from colorful cotton or wool and stand out to the casual observer, focusing exclusively on maguey bags illustrates differences as well. The construction techniques of looping, linking, knitting, and weaving are one way to distinguish them. Other variations are strap construction, strap length, and strap attachment to the bag. Another feature is pita softness and color based on fiber-extraction methods and maguey species. Bags made from fibers extracted using fire or water are yellow in color compared with those made from raw maguey extractions, which are white. The maguey fiber of Concepción is softer than that of Cotzal due to the species harvested, and this is evident in the final product.

Another contrast between looped bags of these two regions is that those of Concepción have a long strap and a loop, whereas Cotzal bags have two straps and no loop. The knit bags of San Marcos la Laguna have one long strap and a short one with a loop at the end. In Chichi there is one long strap and a very short strap with a loop. The final row of "finger-knit" bags from San Marcos is figure-eight looped, whereas the final row in Chichi bags is cast off, as in "true" knitting. The knit maguey bags found in the Sololá and El Quiché regions closely resemble the wool ones of the same areas. The actual method for manipulating the knitting needles is different in each location but most likely originated from the same concept.

The linked bags of San Andrés Huista are unique to the country of Guatemala, but only because of the population's migration from a larger area where linking is prevalent to a locale where looping dominates. Linking is common in the Mexican state of Chiapas, and I am unsure of its history there.

Bags made in the central (San Juan Sacatepéquez) and eastern (Comapa and Olopa) highlands are constructed from a warp-faced, plain-woven cloth. Loom-woven bags are more varied than looped ones. The basic square woven bag is made leaving warp ends that are wrapped and become the handles, in part because it is difficult to weave right up to the ends of the warp. Sembradores have one selvaged side and the other is left long, finished with braids. Caristas and glincas are introduced styles and use the same type of woven fabric as do standard bags. The fancy bags and saddlebags of San Juan Sacatepéquez require greater weaving skills than do the standard bags made in the East. This is unsurprising because women of this municipality are superb backstrap weavers and maguey weavers probably absorbed their techniques.

Looped morrales are made in both regions and could also be the traditional bag style, with woven bags (easier and much quicker to make) a later development, although, in areas where backstrap looms are used for weaving traje, it seems that looped or knit bags are more common. This could be because usually it is the men who make and use the bags, as is the custom, and they probably determined early on that looping was easier and more portable.

I believe that the Cobán macramé bags in the Museo Ixchel's collection are uncommon and that most likely a tourist or development worker introduced the technique to maguey workers. Cross-knit looped bags are also rare because they are labor intensive and require much maguey.

The visual language of the net bag extends beyond Guatemala—in towns where tourists congregate I often see maguey bags from Mexico, Panama, Columbia, Paraguay, El Salvador, and Ecuador. Each one is noticeably different in shape, pattern, and technique and thus reveals its origin to those who know the language.

Appendix 2

Bags, from Beginning to End

Looped and Knit

The beginning point for these bags is the bottom corner. The first row determines the width of the bag. The same strand spirals around until the final height is reached. The final row is doubled or cast off. The bags are usually square, about twelve by twelve inches.

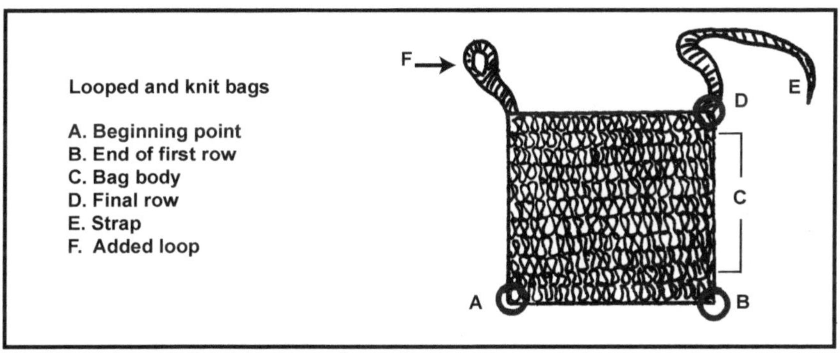

Figure A.2.a

Linked

Linked bags are made sideways. The row of loops placed on the "loom" determines the height of the bag, and the number of rows that are linked determines the width. One side is made and then it is removed from the "loom" and connected to the second side as that side is constructed. After the completed bag is removed from the "loom" the open side is sewn together.

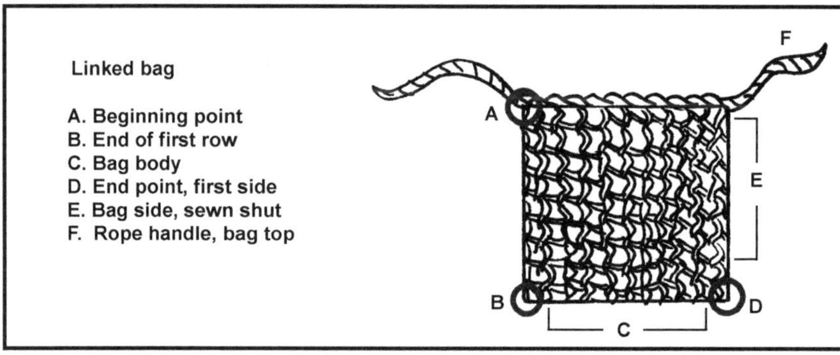

Figure A.2.b

Woven

Woven panels are folded in half and sewn together with a strand of maguey to form the body of the bag. The three to four inches of unwoven warp ends are braided into ten strands on either side and then tied together. A long cord is attached to the braids, which forms the two handles.

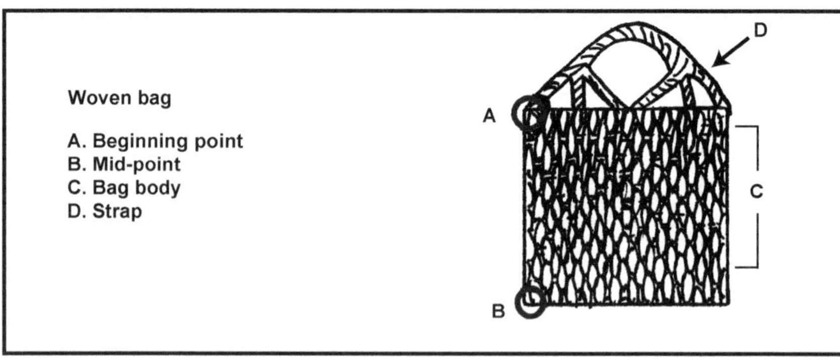

Figure A.2.c

Appendix 3

Fancy Saddlebag Components

The fancy saddlebag is one of the more complicated maguey products and combines weaving, plaiting, braiding, and sprang.

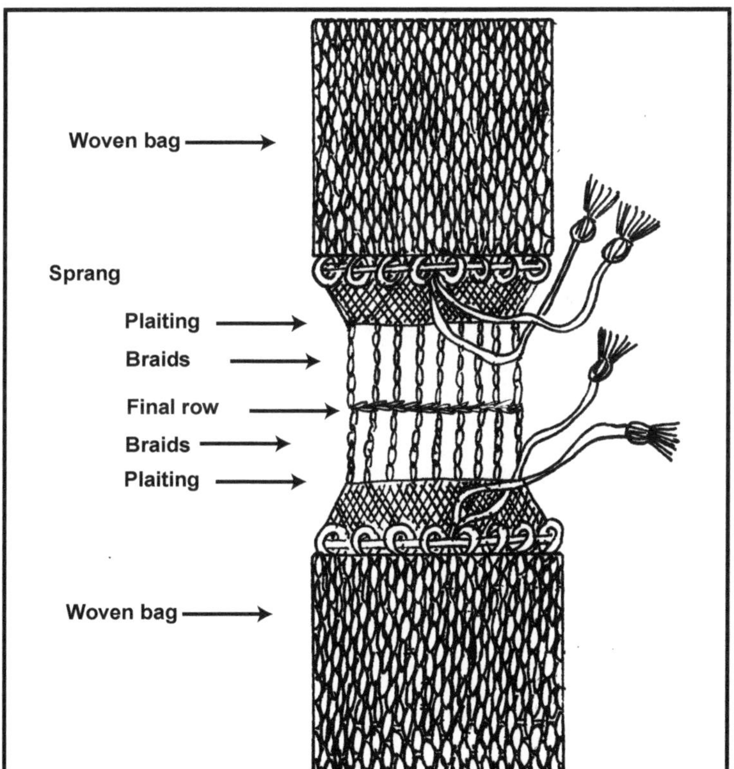

Figure A.3 Fancy saddlebag components.

Appendix 4

Mayan Terminology

Table A4: Mayan Terminology

English	Spanish	K'iche'	Achi	Popti'	Q'eqchi'	Ixil	Mam
maguey	maguey	ki	sa:ki'	sajchi	ik'e	sajchee	ch'ut, chjoj
spun maguey	pita	k'a'm	k'a'aam	tx'anh	ch'ina k'aham	k'a'j	ky'jaj
rope	lazo	kolob'	nimk'a'aam	tx'anh	k'aham	ijvil	aqwil
net bag	morral	chim	teb'	moral	ch'ina champa	sam	pa q'ob'pa
net bag	matate	k'a'm chim	chiim	tz'ispa	champa	k'ach sam	tq'ob'pa
cargo net	red	k'at	k'aat	hak'	soq'	k'ach	xyub' pa
hammock	hamaca	ab'	aab'	atx'b'anh	ab'	ab'	utz
tumpline	mecapal	patän	patan	popom	taab'	xe'k	q'ob'xeky
halter	gamarrón	ukolob' uplaj kej	uchi'ximab'al	xhakima	xaqab'aal	xhekeema	k'alwichej
crupper	tenedor	upas uje' kej	jekebal eqa'n	yet snheh cheh			
bag	bolsa	chim	teb'	tx'uy	ch'ina jelool	tx'u'y	saqchiy ch'u'ẍ
saddlebag	alforja	eqab'al wa	jek'ab'al		taqleb'		
pad	enjalma	utesb'al kej	alb'arto	xhalma	chunleb'		jerk
sack	costal	koxtar	wiriin	koxhtal	koxtal	saako	ch'u'ẍ
cinch	cinta	xaq'ab'	kolob'-xoka	sintahe	tz'ul-leb'	siinta	si'pj
firework	bomba	poq'aj	ja'b'al		moqleb'		q'aq'

Table A4: Mayan Terminology *(cont.)*

English	Spanish	Kaqchikel	Tz'utujil	Poqomám	Ch'orti'	Poqomchi'
maguey	maguey	saqkïy	sajkiij	saqkiy	waka'	sajkih
spun maguey	pita	k'am	k'aam	k'ajaam	jaxb'ir sukchij	k'ahaam
rope	lazo	kolo'	kolo'	ihqul	sumb'iru't	nimaq k'ahaam
net bag	morral	peqës	ya'l	q'eya'l	k'atar chijr	chiim
net bag	matate	peqës	ya'l	ya'l	tz'unpum	chiim
cargo net	red	ya'l	k'amaya'l	ya'l	noj chijr	ya'l
hammock	hamaca	silon k'am	warab'al k'aam	waralk'ajaam	ab'	aab'
tumpline	mecapal	tapäl-patän	teem	q'ab'ihq	ta'b'	pitan, patnaal
halter	gamarrón	b'osal	b'osaal	rupunit keej		
crupper	tenedor	jik'öy k'am		ihqul		
bag	bolsa	peqës	ya'l	jamb'al	chuchu' mukuk	kaxlan chiim
saddlebag	alforja	ruq'ub'al kej	yuqub'el	wi'b'al rihq keej		
pad	enjalma	ruq'ub'al kej	ejqab'al	to'oneel k'olob'al	t'irpatchij	
sack	costal	kostal	koxtaar	koxtaar	mukuk	koxtaal
cinch	cinta	kolo'	paas	siinta	ch'ajnar	b'ak'b'al
firework	bomba	aj		alkapuus	b'u't'k'ajk	

Source: The Mayan terminology was personally provided to me by various translators for K'ulb'il Yol Twitz Paxil in 2008.

Appendix 5

Scientific Names for Cultivated Maguey[a]

Table A5: Scientific Names for Cultivated Maguey[a]

Municipality	Common Name	Scientific Name[b]	Fiber-Extraction Method
Concepción Tutuapa	mecate	*Beschorneria albiflora* Matuda	steamed
San Andrés Huista	maguey de oriente (Garcia-Mendoz 2006, 182)	*Agave angustifolia* Haw. var. *nivea* (Trel.)	retted
San Juan Cotzal	maguey	*Furcraea aff. guatemalensis* Trel.	roasted
Santa Lucía la Reforma	maguey	*Agave* aff. *angustifolia* Haw.	roasted
Chichicastenango	ch'ech, chech or chiyute, chijute	*Furcraea quicheensis* Trel.	steamed
San Pablo la Laguna	maguey	*Agave americana* L.	retted
San Marcos la Laguna	maguey	*Agave* aff. *americana* L.	retted
Acatenango	maguey de oriente (García-Mendoza 2006, 182)	*Agave angustifolia* Haw. var. *nivea* (Trel.)	raw
San Juan Sacatepéquez	maguey sembrado	*Furcraea* sp. nov.	retted
Comapa	maguey verde	*Agave angustifolia* Haw. var. *letonae*	raw
	maguey blanco	*Agave americana* L.	raw
Jocotán	maguey de castillo (verde)	*Agave angustifolia* Haw. var. *letonae*	raw
	maguey de seda (blanco)	*Agave americana* L.	raw
	maguey sin espina or maguey extranjero	*Agave sisalana* Perr.	
Olopa	mescal	*Furcraea* sp. nov.	roasted
San Juan Chamelco	maguey	*Furcraea* aff. *guatemalensis* Trel.	roasted

[a] These are the species that I saw being processed in Guatemala. Other maguey species used for their fiber in Guatemala are *Agave angustifolia* Haw. var. *angustifolia*, *Agave fourcroydes* Lem., *Furcraea tuberosa* (Mill.), and *Yucca elephantipes* Regal (García-Mendoza 2006, 182).

[b] The scientific names were personally provided to me by García-Mendoza, except for *Furcraea quicheensis* Trel. (Camposeco Mateo 1994, 29).

Glossary

agave	maguey
aldea	village or hamlet
alforja	(argana or arganilla [eastern term]) saddlebag
aliso	alder; the cambium (inner bark) is used for a red-brown dye
Altiplano	high plateau; western highlands
AMO	Asociación de Mujeres Olopenses
Arawak	South American tribe
ayudante	bus assistant
backstrap loom	loom made of sticks, with a strap that goes around the weaver's back to control tension
bolsa de mania	eastern term for a loom-woven bag
bomba	pyrotechnic (firework)
cabo	tool for scraping maguey (San Juan Cotzal)
cachelpa	matate (Concepción Tutuapa)
carista	rectangular bag (Olopa)
carrizo	wicker or cane for basket making
che'ch	(chijute) *Furcraea quicheensis* Trel.
cinta	(cincha) cinch: strap that goes under a horse's belly; loom-woven or ply-split darned
cirenuela	small, handheld spinning wheel (torno de mano)
cohetería	fireworks shop
conacaste	taravilla wood (*Enterolobium cyclocarpum*)
corte	woman's wraparound "skirt"
costal	sack
finger-knitting	adaptation of knitting, where the fingers move loops over three sticks (San Marcos la Laguna)

fixed-vertical loom	loom identical to a backstrap loom, except it is vertical and pegged into the ground (San Juan Sacatepéquez)
gamarrón	horse halter (eastern term)
glinca	round bag (Olopa)
guachipilín	taravilla and spinning-wheel wood (*Diphysa robinioides*)
hamaca	hammock
henequén	*Agave fourcroydes*
horizontal loom	a loom consisting of either two forked sticks or four sticks "planted" in the ground, with crossbeams, so that the warps lie horizontally for weaving (Olopa)
huipil	Maya woman's blouse
interlacing	single-element "weave" used to make bag straps
ixtle	Nahuatl term for unspun maguey fibers
jáquima	horse halter (western term)
jarcia	term for working with maguey
knitting	single-element technique using needles
la brea	warm pine pitch for making bombas
Ladino	a person of mixed blood who speaks Spanish, follows Spanish customs, and wears European-style clothing
lazo	rope
linking	technique commonly used for making hammocks and bags in Chiapas, Mexico, and San Andrés Huista, Huehuetenango; can be a single-element technique or two single-element technique
looping	single-element technique; variations include simple, figure-eight, cross-knit, and single interconnected
machucador	pounding tool; wood mallet (San Pablo and San Marcos la Laguna)
mania de palo	scraping tool (Comapa)
masa	cornmeal
matate	a small bag made with linking or figure-eight and worn-tumpline method, with the strap across the user's forehead and the bag resting on the back; looser construction than a morral
mecapal	tumpline
mecate	*Beschorneria albiflora* Matuda (Concepción Tutuapa)
milpa	cornfield
mordant	dye fixative
morral	small bag made with looping, knit, or crochet techniques
Nahuatl	Aztec language

Occidente	western highlands
Oriente	eastern highlands
palo	three-legged, five-inch-wide or so length of wood with a piece of leather, smooth piece of tire, inner tube, or old shoe bottom attached to the top; serves as a "leg" by setting it on the ground next to the spinner or on their lap
parenquima	(sacador) term used by Osborne (1965, 175) for the tool used to separate the fibers from the pulp of the maguey leaf
penca	maguey leaf
perraje	(rebozo) shawl
petate	mat; woven from lake reeds or palm leaves
pita	spun maguey fiber
pita floja	soft unspun fiber from the *Bromelia* plant; also used by Osborne (1965, 248) to refer to unspun maguey fibers
plain weave	two element (warp and weft); also called tabby, where each weft passes alternately over and under successive warp units (Emery 1966, 76)
ply-split braiding	(split-ply) technique where the weft passes diagonally between the ply of the warp
ply-split darning	(split-ply) technique where the weft passes perpendicularly between the ply of the warp
pyramid knot	ornamental knot having a pyramid shape; the basis for bag straps (Graumont and Hensel 1970, 320–321)
quintal	approximately one hundred pounds
rebozo	(perraje) shawl
red	cargo net
rodillera	"kilt" worn by men in certain areas (San Marcos la Laguna)
ronrón	common name for beetle or weevil that eats maguey plants
sacador	(parenquima) term used for the tool used to separate the fibers from the pulp of the maguey leaf
sak-chi	Mayan for maguey (all species except *Furcraea quicheensis* Trel. and *Beschorneria albiflora* Matuda)
selvage	finished edge
sembrador	a small loom-woven bag used for holding corn kernels during planting (San Juan Sacatepéquez)
servilleta	small piece of woven fabric used for carrying food and other items

sisal	*Agave sisalana*
sivadera	simple looped bag (Concepción Tutuapa)
S-twist	strands spun in a clockwise direction
taravilla	(tarabilla) rope maker
tenedor	(crupper) a strap that goes around a pack animal's tail and steadies the cargo holder
torno de mano	small, handheld spinning wheel; cirenuela
traje	traditional Maya clothing
trementina de pino	pine pitch, used for spinning
tzute	(s'ut) Mayan word referring to a square or rectangular woven garment cloth
vara	(abrazo) Spanish measurement equaling 33.6 inches
vertical-frame loom	a loom made of two poles that, when in use, lean against the outside of a house at a vertical angle (Comapa)
warp	longitudinal threads connected to the loom
weft	threads that weave in and out of the warp
Z-twist	strands spun in a counterclockwise direction

Bibliography

Aldana, Llanos, and Ma. Elena Valdes Estrada. 2002. "Scyphophorus Acupunctatus (Coleoptera: Curculionidae) Attacking Polianthes Tuberosa (Lilialers: Agavaceae) in Morelos, Mexico." *Florida Entomologist* 85, no. 2, 392–393, Mexico City, Mexico.

Anawalt, Patricia Rieff. 1981. *Indian Clothing before Cortes: Mesoamerican Costumes from the Codices*. Norman: University of Oklahoma Press.

Anderson, Marilyn. 1978. *Guatemalan Textiles Today*. New York: Watson-Guptill.

Artesanía Ch'orti'/Amor Ch'orti'. 2010. Catálogo de Productos. Guatemala City.

Asociación de Mujeres Olopenses. 2004. Brochure. Guatemala City.

———. 2009. *Catalogo de Productos*. Guatemala City.

Baker, Sylvia. 1985. *Make Your Own Bilum*. Brisbane: Boolarong Publications.

Benítez, Fernando. 1962. *Ki: El Drama de un Pueblo y de una Planta* (2d ed.). Mexico City: Fondo de Cultura Económica.

Berger de White, Mónica, and Matilde Ivic de Monterroso. 2008. "En Busca de las Raíces del Maguey en Guatemala." In: *Ciencia y Técnica Maya: con la coloboracián del Consejo Nacional de Ancianos Principales y Guías Espirituales Maya, Xincas y Garífunas*, pp. 135–183. Guatemala City: Fundación Solar.

Broudy, Eric. 1979. *The Book of Looms: A History of the Handloom from Ancient Times to the Present*. Hanover and London: University Press of New England.

Camposeco Mateo, José Balvino. 1994. *An Chech, Sajchi Oki: El Maguey y Sus Usos en Guatemala*. Cleveland: Yax te' Foundation Books.

Centro Communitario Educativo Pokomchi'. 2001. *Hostal Nuboso Maya-Pokomchi Pamuquense* (brochure). Guatemala City.

Cojolya Association of Mayan Women Weavers. 2010. Web site: www.cojolya.org.

Collingwood, Peter. 1974. *The Techniques of Sprang, Plaiting on Stretched Threads.* New York: Design Books, Watson-Guptill.

———. 1998. *The Techniques of Ply-Split Braiding.* Petaluma, Calif.: Unicorn Books and Crafts.

Dary, Claudia, and Aracely Esquivel. 1983. "Las Labores de Jarcia en Comapa, Jutiapa, Guatemala." In: *La Tradicion Popular,* Boletín del Centro de Estudios Folklóricos. Guatemala City: Universidad de San Carlos de Guatemala.

Davidson, D. S. 1935. "Knotless Netting in America and Oceania." *American Anthropologist,* no. 37:117–134.

De Luco, Pam. 2001. "The Chácara of the Ngöbe Women: Handspun String Bags from Panama." *Spin Off* summer:62–65.

Dewey, Lyster H. 1965. *Fibras Vegetales y Su Producción en América* (3d ed.). Mexico City: Centro Regional de Ayuda Técnica.

Domingo, Armando Ortíz. 1990. "Jarcia." In: *Distribución Geográfica de las Artesanías de Guatemala,* pp. 68–88. Guatemala City: Subcentro Regional de Artesanías y Artes Populares.

Drooker, Penelope. 1981. *Hammock Making Techniques.* Sanbornville, N.H: Penelope B. Drooker.

Edholm, Steven, and Tamara Wilder. 1995 (rev.). *Cordage.* Ukiah, Calif.: Steven Edholm and Tamara Wilder.

Emery, Irene. 1966. *The Primary Structure of Fabrics.* Washington, D.C.: The Textile Museum.

England, Nora C. 2001. *Introducción a la Gramática de los Idiomas Mayas.* Guatemala City: Cholsamaj Foundation.

Feldman, Lawrence H. 1985. *A Tumpline Economy: Production and Distribution Systems in Sixteenth-Century Eastern Guatemala.* Culver City, Calif: Labyrinthos.

Fernández, Vicente. 2000. "La Ley del Monte." *Historia de un Ídolo, Vol. 1.* Album label: Sony International, Mexico City.

Fowler, William R., Jr. 1989. *The Cultural Evolution of Ancient Nahua Civilizations: The Pipil-Nicarao of Central America.* Norman: University of Oklahoma, Norman.

García-Mendoza, Abisai. 1992. *Con Sabor a Maguey.* Mexico City: Universidad Nacional Autónoma de Mexico.

———. 2006. "Diversidad, Distribución, e Importancia Económica de las Agavaceae de Gíuatemala." In: *Biodiversidad de Guatemala,* pp. 175–186. Guatemala City: Universidad del Valle de Guatemala.

Gentry, Howard Scott. 1982. *Agaves of Continental North America*. Tucson: University of Arizona Press.

Girard, Rafael. 1949. *Los Chortis ante el Problema Maya: Historia de las Culturas Indígenas de América, desde Su Origen Hasta Hoy*. Mexico City: Antigua Libreria Robredo, Colección Cultura Precolombina.

González, David. 2003. "Justice in Guatemala Can Be Gasoline and a Match." *New York Times*, March 8.

Granberg, Wilbur J. 1970. *People of the Maguey: The Otomi Indians of Mexico*. New York: Praeger.

Graumont, Raoul, and John Hensel. 1970. *Encyclopedia of Knots and Fancy Rope Work* (4th ed.). Cambridge, Md.: Cornell Maritime Press.

Heisse, Susana. 2000. "Tying Tradition with Progress: Rescuing the Ancient Atitlán Craft of Maguey Weaving." *Guatemalan Post* (Guatemala City), July 6–12.

———. 2002a. *Maguey*. Sololá, Guatemala: Sussana Heisse.

———. 2002b. *The Maguey Plant*. Sololá, Guatemala: Sussana Heisse.

———. 2002c. *The Soul Project*. Sololá, Guatemala: Sussana Heisse.

———. 2002d. *The Traveling Maguey*. Sololá, Guatemala: Susana Heisse.

Hopkins, Richard. 2001. "More Early Cord Making Tools." In: *Knotting Matters*, no. 72:20–21. Chester, U.K.: International Guild of Knot Tyers.

Irigoyen, Renan. 1950. *Los Mayas y el Henequén*. Mérida, Mexico: Henequeros de Yucatán.

Johnson, Imgard Weitlaner. 1971. "Basketry and Textiles." In: *Handbook of Middle American Indians*, vol. 10, pp. 297–320. Austin: University of Texas Press.

Kelsey, Vera, and Lilly de Jongh Osborne. 1961. *Four Keys to Guatemala*. New York: Funk & Wagnalls.

K'ulb'il Yol Twitz Paxil. 2001. *Toponimias Maya Jakalteka*. Jacaltenango, Guatemala: Academia de Lenguas Mayas de Guatemala.

Lothrop, Samuel K. 1929. "The Henequen Industry of San Pablo, Guatemala." *Indian Notes*, no. 6:120–129. Washington D.C.: Museum of the American Indian, Heye Foundation.

MacKenzie, Maureen. 1991. *Androgynous Objects: String Bags and Gender in Central New Guinea*. Amsterdam: Harwood Academic Publishers.

MacLeod, Murdo. 1973. *Spanish Central America: A Socioeconomic History, 1520–1720*. Berkeley: University of California Press.

MacVean, Ana Lucrecia de. 2006. *Plantas Útiles de Sololá, Guatemala*. Guatemala City: Universidad del Valle de Guatemala.

Magaña, Silvia de. 2001. "Con Fibra de Maguey." *Prensa Libre* (Guatemala City), August 5.

Mauersberger, Herbert R. 1947. *Matthews' Textile Fibers: Their Physical, Microscopical, and Chemical Properties* (5th ed.). New York: John Wiley and Sons.
McBryde, Felix Webster. 1945. *Cultural and Historical Geography of Southwest Guatemala*. Washington, D.C.: Smithsonian Institution, Institution of Social Anthropology, no. 4.
Metz, Brent. 2006. *Ch'orti'-Maya Survival in Eastern Guatemala: Indigeneity in Transition*. Albuquerque: University of New Mexico Press.
Museo de Café. 2007. From a sign accompanying an exhibit at the museum, in Centro la Azotea, Jocotenango, Guatemala.
Nash, Manning. 1967. "Indian Economics." In: *Handbook of Middle American Indians*, vol. 6, pp. 87–102. Austin: University of Texas Press.
Nordland, Odd. 1961. *Primitive Scandinavian Textiles in Knotless Netting*. Oslo: Oslo University Press.
O'Neale, Lila M. 1934. Peruvian "Needleknitting." *American Anthropologist* 36, no. 3:405–430.
———. 1945. *Textiles of Highland Guatemala*. Washington, D.C.: Carnegie Institute of Washington.
Osborne, Lilly de Jongh. 1935. *Guatemalan Textiles*. New Orleans: Middle American Research Institute, Publication 6.
———. 1956. *Four Keys to El Salvador*. New York: Funk & Wagnalls.
———. 1965. *Indian Crafts of Guatemala and El Salvador*. Norman: University of Oklahoma Press.
Parsons, Jeffrey R., and Mary H. Parsons. 1990. *Maguey Utilization in Highland Central Mexico: An Archeological Ethnography*. Ann Arbor: University of Michigan Press.
Proyecto Maya Chorti. 2002. Catalog. Guatemala City.
Recinos, Adrian, and Delia Goetz. 1953. *The Annals of the Cakchiquel*. Norman: University of Oklahoma Press.
Reina, Ruben E. 1969. "Eastern Guatemalan Highlands: The Pokomames and Chorti." In: *Handbook of Middle American Indians*, vol. 7, pp. 101–132. Austin: University of Texas Press.
Rodman, Debra Hain. 2006. "Gender, Migration, and Transnational Identities: Maya and Ladino Relations in Eastern Guatemala." PhD dissertation, University of Florida, Gainesville.
Rodríguez Rouanet, Francisco. 1990. "Coheteria." In: *Distribución Geográfica de las Artesanías de Guatemala*, pp. 238–257. Guatemala City: Subcentro Regional de Artesanías y Artes Populares.
Rutt, Richard. 1987. *A History of Handknitting*. London: B. T. Batsford.

Salinas Pedraza, Jesús. 2000. "Testimony of an Otomi," *Maguey Artes de Mexico*. No. 51:86–90.
Schery, Robert W. 1972. *Plants for Man*. Englewood Cliffs, N.J.: Prentice Hall.
Schevill, Margot Blum. 1993. *Maya Textiles of Guatemala*. Austin: University of Texas Press.
Seiler-Baldinger, Annemarie. 1994. *Textiles: A Classification of Techniques*. Washington, D.C.: Smithsonian Institution Press.
Sentence, Bryan. 2001. *Art of the Basket: Traditional Basketry from Around the World*. London: Thames & Hudson.
Sheets, Payson (ed.). 2002. *Before the Volcano Erupted: The Ancient Cerén Village in Central America*. Austin: University of Texas Press.
Squier, E. G. (Ephraim George). 1861. *Tropical Fibers: Their Production and Economic Extraction*. New York: Scribner and Company.
Standley, Paul C., and Julian A. Steyermark. 1952. *Flora of Guatemala*. Chicago: Field Museum of Natural History.
Valenzuela-Zapata, Ana, and Gary Paul Nabhan. 2003. *Tequila: A Natural and Cultural History*. Tucson: University of Arizona Press.
Ventura, Carol Ann. 1987. "Sisal Production in Highland Guatemala." *Fiberarts* 14, no. 2:44–45.
———. 1996. *Mayan Hair Sashes Backstrap-Woven in Jacaltenango, Guatemala*. Cleaveland: Yaxte' Foundation Books.
Villatoro, Jose Luis. (n.d.) *Concepción Tutuapa el Techo de Central America*. (n.p.)
Wagley, Charles. 1969. "The Maya of Northwestern Guatemala." In: *Handbook of Middle American Indians*, vol. 7, pp. 46–68. Austin: University of Texas Press.
White, C. Langdon, and Edwin J. Foscue. 1939. "Henequin: The Green Gold of Yucatan." *Journal of Geography* 38:151–155.
Wipplinger, Michele. 1996. *Tintes Naturales para Artesanoa de las Américas*. Washington, D.C.: Organization of American States.
Wisdom, Charles. 1940. *The Chorti Indians of Guatemala*. Chicago: University of Chicago Press.

Further Reading

Adams, Richard N. 1976. *Cultural Surveys of Panama, Nicaragua, Guatemala, El Salvador and Honduras*. Detroit: Blaine Ethridge Books.

Benítez, Fernando. 2000. "Lord Maguey." *Maguey: Artes de Mexico*, no. 51:82–84.

Ceron, Isaias Mendoza, and Una Conger. 1993. *In Tequil de Morrales: Working with Maguey*. Copenhagen: Bianca Luno, A/S.

Collingwood, Peter. 1987. *The Maker's Hand: A Close Look at Textile Structures*. London: Bellew Publishing.

Davidson, D. S. 1933–1934. "Australian Netting and Basketry Techniques." *Journal of the Polynesian Society*, no. 42–43:257–299.

D'Harcourt, Raoul. 1962. *Textiles of Ancient Peru and Their Techniques* (edited by Grace G. Denny and Carolyn M. Osborne; translated by Sadie Brown). Seattle: University of Washington Press.

Dickey, Saundra F. 1964. *A Historical Review of Knotless Netting in South America*. Berkeley: University of California.

Feinman, Gary M., Linda M. Nicholas, and Helen R. Haines. 2002. "Mexico's Wonder Plant." *Archaeology* 55, no. 2:33–35.

Foster, Alice. 1930. "Sisal Production in Semi-Arid Karst Country of Yucatan." *Journal of Geography* 29:16–25.

Hendon, Julia A. 1999. "Spinning and Weaving in Pre-Hispanic Mesoamerica: The Technology and Social Relations of Textile Production." In: *Mayan Clothing and Weaving through the Ages*. Guatemala City: Museo Ixchel del Traje Indígena de Guatemala.

Hopkins, Richard. 2000. "Maya Ropemaking." *Knotting Matters*, no. 69:20–21. Chester, U.K.: International Guild of Knot Tyers.

Kata, Cathy. 2007. *CK Bilum Wear*. Goroka, Papua New Guinea: PJKata Publications.

Markarian, Janet. 1984. "Shigra making in Ecuador." *American Craft* vol. 44, no.1:12–16.
Maudslay, Anne Cary, and Alfred Percival. 1899. *A Glimpse at Guatemala*. London: John Murray.
McCreery, David. 1994. *Rural Guatemala, 1760–1940*. Palo Alto: Stanford University Press.
Oakes, Maud. 1951. *Beyond the Windy Place: Life in the Guatemalan Highlands*. New York: Farrar, Straus and Young.
Olson, James S. 1991. *The Indians of Central and South America: An Ethnohistorical Dictionary*. Westport, Conn.: Greenwood Press.
Osers, David Altalef. 1971. *Centro de Mercado Artesanías*. Guatemala City: Universidad de San Carlos de Guatemala.
Paganini, Ricardo Casteneda. 1959. "La Cultura Tolteca-Pipil de Guatemala." Guatemala City: Editorial del Ministerio de Educación Pública "José de Pineda Ibarra."
Pettersen, Carmen L. 1976. *The Maya of Guatemala: Their Life and Dress*. Guatemala City: Museo Ixchel del Traje Indígena.
Quick, Betsy D., and Judith A. Stein. 1982. *Split-Ply Camel Girths of West India*. Los Angeles: University of California Press.
Rodríguez Rouanet, Francisco. 1985. *Estudio de las Artesanías Populares de Guatemala*. Guatemala City: Subcentro Regional de Artesanías y Artes Populares.
———. 1989. *Artesanías y Producción Artesanal*. Guatemala City: Subcentro Regional de Artesanías y Artes Populares.
Rojas, Alfonso Villa. 1969. "Maya Lowlands: The Chontal, Chol, and Kekchi." In: *Handbook of Middle American Indians*, vol. 7, pp. 230–243. Austin: University of Texas Press.
Rojas, Ulises. 1936. *Elementos de Botánica General*. Guatemala City: Tipográfica Nacional (1925–1936).
Rolden, Denys. 2003. "Productos de Maguey: Las Hamacas son Artesanías de la Región Chorti, Zona de Pobreza." *Prensa Libre* (Guatemala City), May 18.
Roth, H. Ling. 1977. *Studies in Primitive Looms*. McMinnville, Ore.: Robin and Russ Handweavers (reprint).
Roth, Walter E. 1901. "String and Other Forms of Strand: Basketry, Woven Bag, and Net Work." *North Queensland Ethnography: Bulletin*, no. 1. Brisbane: Government Printer.
Sapper, Karl. 2000. *Early Scholars' Visits to Central America: Reports by Karl Sapper, Walter Lehmann, and Franz Termer* (translated by Theodore E. Gutman; edited by Marilyn Beaudry-Corbett and Ellen T. Hardy). Los Angeles:

University of California Press: Cotsen Institute of Archaeology, Occasional Paper 18.

Schevill, Margot Blum (edited with the assistance of Linda Asturias de Barrios; Jeffrey J. Foxx, photographer). 1997. *The Maya Textile Tradition*. New York: Harry N. Abrams.

Schlesinger, Victoria. 2001. *Animals and Plants of the Ancient Maya: A Guide*. Austin: University of Texas Press.

Solis-Aguilar, Juan Fernando, Hector Gonzales-Hernández, Jorge Luis Leyva-Vázquez, Armando Equihua-Martínez, Francisco Javier Flores-Mendoza, and Ángel Martínez-Garza. 2001. "2001 *Scyphophorus acupunctatus* Gyllenhal, Plaga del Agave Tequilero en Jalisco, Mexico." *Agrociencia* 35, no. 6 (Nov./Dec.):663–670.

Stephens, John L. 1969. *Incidents of Travel in Central America, Chiapas, and Yucatan*, vols. 1 and 2. New York: Dover.

Thomas, Ernest S. 1926. "Netting without a Knot." *Man* (Royal Anthropological Institute of Great Britain and Ireland) 26:8–10.

Traven, B. 1993. "Assembly Line." In: *The Night Visitor and Other Stories*, pp. 73–88. Chicago: Ivan R. Dee.

Vargas, Margarita Ramirez. 1995. "Del Maguey al Lazo." *Prensa Libre* (Guatemala City), June 10.

West, Robert C., and John P. Augelli. 1976. *Middle America: Its Lands and Peoples*. Englewood Cliffs, N.J.: Prentice Hall.

Wipplinger, Michele. 2005. *Natural Dye Instruction Booklet*. Seattle: Earthues.

Index

About the Author

Kathryn Rousso began researching maguey textiles while developing her own fiber art, some items of which are made with the same techniques. In 2001, shortly after completing an MFA in Textile Arts and Costume Design at the University of California, Davis, she received a Fulbright grant to conduct fieldwork in Guatemala.

Her first exposure to Maya textiles was in the late 1980s, when, as a Peace Corps volunteer in a Kaqchikel village, she learned backstrap weaving. In addition to her interest in Guatemalan (and Northwest Coast) textiles, she has conducted research on basketry and textiles in Nusa Tenggara, Indonesia; Andean South America; and "Maori" New Zealand. Rousso has a BA from the University of Washington and a Certificate of Merit in Northwest Coast Basketry and Textiles from the Totem Heritage Center in Ketchikan, Alaska.

Her ongoing research is intertwined with weaving, exhibit curating, workshop instruction, presentations, and writings about maguey and related topics. She contributes occasional articles to *Revue*, an English-language magazine published monthly in Guatemala, and wrote the chapters "Morrales" and "Dress of Eastern Guatemala" for the *Berg Encyclopedia of World Dress and Fashion*, vol. 2: *Latin America and the Caribbean*. Her article "Betel Nuts, Beads, and Beliefs in Atoni Society" appeared in *Arts of Asia* magazine.

Rousso's contemporary baskets combine traditional hand-constructed techniques and natural materials from southeastern Alaska and Guatemala, bringing these two worlds together. Future plans include continued net-bag and basket research in Latin America, Southeast Asia, Alaska, and the Pacific.

From October 2005 to November 2009 Rousso resided part time in Mendocino, California, where she was affiliated with the Mendocino Art Center. She currently splits her time between Ketchikan, Alaska, and Guatemala.